The Open University

S342
Science: a third level course

Physical Chemistry
PRINCIPLES OF CHEMICAL CHANGE

BLOCK 1
SCOPE AND LIMITATIONS OF THE THERMODYNAMIC APPROACH

THE S342 COURSE TEAM

CHAIR AND GENERAL EDITOR
Kiki Warr

AUTHORS
Keith Bolton (Block 8; Topic Study 3)
Angela Chapman (Block 4)
Eleanor Crabb (Block 5; Topic Study 2)
Charlie Harding (Block 6; Topic Study 2)
Clive McKee (Block 6)
Michael Mortimer (Blocks 2, 3 and 5)
Kiki Warr (Blocks 1, 4, 7 and 8; Topic Study 1)
Ruth Williams (Block 3)

Other authors whose previous S342 contribution has been of considerable value in the preparation of this Course

Lesley Smart (Block 6)
Peter Taylor (Blocks 3 and 4)
Dr J. M. West (University of Sheffield, Topic Study 3)

COURSE MANAGER
Mike Bullivant

EDITORS
Ian Nuttall
Dick Sharp

BBC
David Jackson
Ian Thomas

GRAPHIC DESIGN
Debbie Crouch (Designer)
Howard Taylor (Graphic Artist)

COURSE READER
Dr Clive McKee

COURSE ASSESSOR
Professor P. G. Ashmore (original course)
Dr David Whan (revised course)

SECRETARIAL SUPPORT
Debbie Gingell (Course Secretary)
Jenny Burrage
Margaret Careford
Shirley Foster
Sue Hegarty

The Open University, Walton Hall, Milton Keynes, MK7 6AA

Copyright © 1996 The Open University. First published 1996. Reprinted 2001

All rights reserved. No part of this publication may be reproduced, stored in a retrieval system or transmitted in any form or by any means, without written permission from the publisher or a licence from the Copyright Licensing Agency Limited. Details of such licences (for reprographic reproduction) may be obtained from the Copyright Licensing Agency Ltd of 90 Tottenham Court Road, London, W1P 9HE.

Edited, designed and typeset by The Open University.

Printed in the United Kingdom by Henry Ling Ltd, The Dorset Press, Dorchester DT1 1HD

ISBN 0 7492 51638

This text forms part of an Open University Third Level Course. If you would like a copy of Studying with The Open University, please write to the Central Enquiry Service, PO Box 200, The Open University, Walton Hall, Milton Keynes, MK7 6YZ. If you have not enrolled on the Course and would like to buy this or other Open University material, please write to Open University Educational Enterprises Ltd, 12 Cofferidge Close, Stony Stratford, Milton Keynes, MK11 1BY, United Kingdom.

CONTENTS

1 INTRODUCTION:
REQUIREMENTS FOR REACTION — 5

2 CHEMICAL EQUILIBRIUM:
USING THE THERMODYNAMIC DATABASE — 6

3 THE GIBBS FUNCTION
AND THE EQUILIBRIUM CONSTANT — 9

4 THE TEMPERATURE DEPENDENCE
OF THE EQUILIBRIUM CONSTANT — 10

5 THE EQUILIBRIUM CONSTANT
AND THE EQUILIBRIUM YIELD — 13
   5.1   The equilibrium constant for a gas reaction — 14
   5.2   Calculating the equilibrium yield — 15

6 SYNTHESIS OF AMMONIA:
THE THERMODYNAMIC 'LIMITS OF OPERATION' — 20

SUMMARY OF BLOCK 1 — 23

OBJECTIVES FOR BLOCK 1 — 26

SAQ ANSWERS AND COMMENTS — 27

ANSWER TO EXERCISE — 31

1 INTRODUCTION:
REQUIREMENTS FOR REACTION

S342 is about physical chemistry, but in a half-credit course it is impossible to give a comprehensive account of such a vast body of knowledge. Rather, we decided to select aspects of the subject and to study these in depth. The different aspects are linked together by a central theme that runs through the Course: an examination of the general chemical principles that govern whether, how, and under what conditions, substances will react with one another.

Broadly speaking, two separate conditions must be fulfilled before a chemical reaction can occur. First, the *equilibrium position* must favour the reaction, and second, the reaction must have a detectable *rate*. In any practical context, and especially in the chemical industry where commercial viability must be the ultimate goal, it is crucial that the chosen reaction should proceed at a reasonable rate. But the *theoretical* limit to the advantages gained by improving the rate is set by the attainment of equilibrium. In other words, the position of equilibrium places an upper limit on the yield that can be obtained from a particular reaction under a given set of conditions: the *equilibrium yield*. Questions concerning the equilibrium yield from a reaction are governed by the laws of thermodynamics.

For this reason, we felt it important to begin the Course with a reminder of your background in chemical thermodynamics.* We then take this analysis a stage further, and show you how thermodynamic data can be used to predict what might be termed the 'limits of operation' for a reaction; that is, the equilibrium yield of product to be expected under a range of reaction conditions. To this end, the discussion throughout Block 1 is largely restricted to reactions in the *gas* phase. As you will see shortly, such reactions allow of a particularly simple technique for predicting the equilibrium yield under a given set of conditions.

Further, to highlight more clearly the scope and limitations of this purely thermodynamic approach, we decided to base the discussion around one specific reaction – the synthesis of ammonia from its elements:

$$N_2(g) + 3H_2(g) = 2NH_3(g) \tag{1}$$

This example was not chosen at random. You may recognize it as the basis of the **Haber–Bosch process**, which lies at the heart of the giant nitrogen-based fertilizer industry. So, by choosing this example, we shall be in a good position to compare the results of our thermodynamic analysis with industrial practice.

One final point. Although we shall use reaction 1 to carry the 'story-line', as it were, the ideas developed below are not, of course, restricted to this one example. We shall generalize the more important results as we go along, either in the text, or by way of examples for you to work through yourself.

STUDY COMMENT   Physical chemistry is a quantitative subject. Throughout the Course, we shall often be in the business of developing and handling mathematical expressions that interrelate the physical quantities of interest. For the most part, this does not involve any new mathematical skills: the algebraic manipulations required were introduced in the Science Foundation Course, and developed further in the Second Level Inorganic Course. The best way to build up your confidence with such manipulations is to get as much practice as possible. For example, it would be a good plan to get into the habit of working through the derivations in the text, checking for yourself how we got from one line to the next.

---

* This subject was introduced in S247 *Inorganic Chemistry: Concepts and Case Studies* – the Second Level Course that is a recommended prerequisite for S342. Here, and elsewhere in S342, important concepts that you should be familiar with already are indicated by a reference to the 'Second Level Inorganic Course' – or indeed, to the 'Science Foundation Course' if that is more appropriate.

On a more specific note: there are places in this Block where you will need to use two general skills that will be drawn on repeatedly throughout S342:

1   The ability to manipulate general expressions that contain the logarithm or exponential of a physical quantity.

2   The ability to interpret correctly the label attached to the axis of a graph, or in the heading to a column in a table of data.

Sections 1 and 2 of the AV Booklet, and the accompanying tape sequences (bands 1 and 2 on audiocassette 1) are designed to provide guidance in these areas. Aim to consult this advice as soon as possible, referring to:

- Section 1 of the AV Booklet, if you are uncertain about the meaning of logarithms (both 'to the base ten' and 'to the base e') or exponentials;
- Section 2 of the AV Booklet, if you are unsure about the correct way to head columns of data in a table or to label the axes of a graph.

2 CHEMICAL EQUILIBRIUM:
USING THE THERMODYNAMIC DATABASE

As you may recall from the Second Level Inorganic Course, the laws of thermodynamics allow an elegant restatement of our first condition for reaction – that the equilibrium position must be favourable – in terms of the **standard molar Gibbs function change** (or Gibbs free energy change) for the reaction, written $\Delta G_m^\ominus$.

> There, we took the criterion $\Delta G_m^\ominus < 0$ as the definition of a thermodynamically favourable reaction; conversely, a reaction was considered to be unfavourable if $\Delta G_m^\ominus > 0$.

For our present purposes, the importance of this simple criterion is that values of $\Delta G_m^\ominus$ do not necessarily have to be determined by studying the reaction itself; they can be calculated from the results of experiments on the individual reactants and products involved. Calculations like this open the door to the vast storehouse of thermodynamic data in the chemical literature, and thereby increase enormously the predictive power of the subject.

If you consult Section 2 of the S342 *Data Book*, you will find a compilation of **thermodynamic data** at 298.15 K (25 °C) for a selection of elements, compounds and aqueous ions: it should have a familiar look. It is important that you recall the meaning of the quantities recorded there, and are confident about manipulating them. To refresh your memory, consider our example reaction:

$$N_2(g) + 3H_2(g) = 2NH_3(g) \tag{1}$$

Data for the substances in equation 1 are collected in Table 1. Notice that there are entries for three thermodynamic quantities: $\Delta H_f^\ominus$, $\Delta G_f^\ominus$, and $S^\ominus$. In part, this reflects the fact that values of $\Delta G_m^\ominus$ are often calculated from standard **enthalpy** (H) and **entropy** (S) data because this method has the widest application.

Table 1 Thermodynamic data at 298.15 K for the substances in reaction 1.

Substance	State	$\dfrac{\Delta H_f^\ominus}{\text{kJ mol}^{-1}}$	$\dfrac{\Delta G_f^\ominus}{\text{kJ mol}^{-1}}$	$\dfrac{S^\ominus}{\text{J K}^{-1}\text{ mol}^{-1}}$
N_2	g	0	0	191.6
H_2	g	0	0	130.7
NH_3	g	−46.0	−16.4	192.5

■ Do you recall the relation between $\Delta G_m^\ominus$ on the one hand, and the corresponding values of the **standard molar enthalpy and entropy changes** ($\Delta H_m^\ominus$ and $\Delta S_m^\ominus$, respectively), on the other?

■ This important relation comes from the definition of the Gibbs function ($G = H - TS$), and is given, for a reaction at a constant temperature T (and implicitly at constant pressure), by the expression:

$$\Delta G_m^\ominus = \Delta H_m^\ominus - T\Delta S_m^\ominus \tag{2}$$

■ Now concentrate on the column headed $\Delta H_f^\ominus / \text{kJ mol}^{-1}$ in Table 1. What does the subscript f imply in this context?

■ $\Delta H_f^\ominus$ denotes the **standard enthalpy of formation** of a substance, at 298.15 K and at a standard pressure of 10^5 Pa (more on which later).

To quote the more extended definition given in Section 2 of the S342 *Data Book*:

> $\Delta H_f^\ominus$ is the standard molar enthalpy change for the reaction in which one formula unit of the substance is formed from its elements, each in the form (the *reference state*) that is stable at a temperature of 298.15 K, and at a pressure of 10^5 Pa.

■ Why are the entries in the enthalpy column of Table 1 zero for both $H_2(g)$ and $N_2(g)$?

■ For both of these elements, the diatomic gas *is* the reference state at 298.15 K and 10^5 Pa – effectively room temperature and pressure – so $\Delta H_f^\ominus$ is zero by definition.

The reference states of other elements can easily be identified in data tables because their $\Delta H_f^\ominus$ values must also be zero.

■ Now use information from Table 1 to calculate $\Delta H_m^\ominus$ at 298.15 K for reaction 1.

■ The answer is $\Delta H_m^\ominus = 2\Delta H_f^\ominus(NH_3, g) = -92.0 \text{ kJ mol}^{-1}$. Hopefully, you recognized that equation 1 is just *twice* the formation reaction for $NH_3(g)$:

$$\tfrac{1}{2}N_2(g) + \tfrac{3}{2}H_2(g) = NH_3(g) \tag{3}$$

More generally, for any chemical reaction written in 'alphabetical form' as follows:

$$a\text{A} + b\text{B} + \ldots = p\text{P} + q\text{Q} + \ldots \tag{4}$$

the value of $\Delta H_m^\ominus$ can be calculated from the expression:

$$\Delta H_m^\ominus = \{p\Delta H_f^\ominus(\text{P}) + q\Delta H_f^\ominus(\text{Q}) + \ldots\} - \{a\Delta H_f^\ominus(\text{A}) + b\Delta H_f^\ominus(\text{B}) + \ldots\} \tag{5}$$

In just the same way, values of $\Delta S_m^\ominus$ can be calculated from the **absolute entropies**, $S^\ominus$, of the substances involved in the reaction. For reaction 1, for example:

$$\Delta S_m^\ominus = 2S^\ominus(NH_3, g) - S^\ominus(N_2, g) - 3S^\ominus(H_2, g)$$
$$= \{(2 \times 192.5) - 191.6 - (3 \times 130.7)\} \, J\,K^{-1}\,mol^{-1}$$
$$= -198.7 \, J\,K^{-1}\,mol^{-1}$$

■ What, then, is the standard entropy of formation of $NH_3(g)$ at 298.15 K?

■ Since equation 1 is twice the formation reaction for $NH_3(g)$ (equation 3), $\Delta S_f^\ominus(NH_3, g) = \tfrac{1}{2}\Delta S_m^\ominus(1) = -99.4 \, J\,K^{-1}\,mol^{-1}$.

The fact that this value of $\Delta S_f^\ominus$ does *not* feature in Table 1 provides a timely reminder. Here – and throughout the more extensive tables in the S342 *Data Book* – the entries in the entropy column are the *absolute* entropies, $S^\ominus$, of the substances, *not* their entropies of formation, $\Delta S_f^\ominus$. There are certain kinds of calculations (SAQ 2 at the end of this Section provides two common examples) where the distinction between $S^\ominus$ and $\Delta S_f^\ominus$ is of crucial importance.

■ Returning to the reaction in equation 1, we have calculated that, at 298.15 K, $\Delta H_m^\ominus = -92.0 \, kJ\,mol^{-1}$ and $\Delta S_m^\ominus = -198.7 \, J\,K^{-1}\,mol^{-1}$. Now use equation 2 to determine the value of $\Delta G_m^\ominus$ at 298.15 K.

■ The answer is $-32.8 \, kJ\,mol^{-1}$. The figure is obtained by substituting in equation 2 (with $T = 298.15$ K), remembering to take account of the fact that $\Delta S_m^\ominus$ is quoted in units of $J\,K^{-1}\,mol^{-1}$, *not* $kJ\,K^{-1}\,mol^{-1}$. Thus,

$$\Delta G_m^\ominus = (-92.0 \, kJ\,mol^{-1}) - (298.15 \, K) \times (-198.7 \times 10^{-3} \, kJ\,K^{-1}\,mol^{-1})$$
$$= (-92.0 + 59.24) \, kJ\,mol^{-1}$$
$$= -32.8 \, kJ\,mol^{-1}$$

■ Is this answer consistent with the entries in the column headed $\Delta G_f^\ominus / kJ\,mol^{-1}$ in Table 1?

■ Yes. $\Delta G_f^\ominus$ is the value of $\Delta G_m^\ominus$ for the same formation reaction to which $\Delta H_f^\ominus$ refers. Thus, for reaction 1, $\Delta G_m^\ominus = 2 \Delta G_f^\ominus (NH_3, g)$, as above.

STUDY COMMENT If you would like more practice with using the thermodynamic database at this stage, try the following SAQs. SAQ 1 concentrates on gaseous reactions—the type of reaction of most immediate concern in this Block. SAQ 2 is somewhat broader, including reactions that involve solids, liquids and aqueous ions as well: make sure you consult the notes at the beginning of Section 2 in the S342 *Data Book* before tackling it.

SAQ 1 (revision) Use information from the S342 *Data Book* to calculate $\Delta H_m^\ominus$, $\Delta S_m^\ominus$, and hence (using equation 2) $\Delta G_m^\ominus$ at 298.15 K for each of the following reactions:

(a) $CH_3OH(g) + \tfrac{1}{2}O_2(g) = HCHO(g) + H_2O(g)$ (6)

(b) $2N_2O_5(g) = 4NO_2(g) + O_2(g)$ (7)

(c) $2CO(g) + 3H_2(g) = C_2H_2(g) + 2H_2O(g)$ (8)

SAQ 2 (revision) This question is concerned with the information in Table 2. Taking any further information you require from the S342 *Data Book*, fill in the two blank entries in Table 2.

Table 2 Thermodynamic data at 298.15 K for scandium (Sc).

Substance	$\dfrac{\Delta H_f^\ominus}{\text{kJ mol}^{-1}}$	$\dfrac{\Delta G_f^\ominus}{\text{kJ mol}^{-1}}$	$\dfrac{S^\ominus}{\text{J K}^{-1}\text{ mol}^{-1}}$
Sc(s)	0	0	34.6
Sc^{3+}(aq)		−586.6	−255.2
ScBr$_3$(s)	−743.1		167.4

3 THE GIBBS FUNCTION
AND THE EQUILIBRIUM CONSTANT

STUDY COMMENT If you have not already done so, this would be a good point to consult Section 1 of the AV Booklet, and listen to the accompanying tape sequence (band 1 on audiocassette 1).

To return to the synthesis of ammonia:

$$N_2(g) + 3H_2(g) = 2NH_3(g) \tag{1}$$

Remember that our aim is to examine how far the thermodynamic quantities calculated above allow a rationalization (on both a qualitative *and* a quantitative level) of the operating conditions used in practice to produce ammonia in the Haber–Bosch process.

To begin with the value of $\Delta G_m^\ominus$: for reaction 1, $\Delta G_m^\ominus < 0$ at 298.15 K, so the synthesis of ammonia from its elements certainly looks feasible on thermodynamic grounds at this temperature. But what about the actual *yield* of ammonia when equilibrium is attained? Can we expect effectively complete conversion into the desired product? If not, how can the equilibrium position be characterized more precisely?

In the Second Level Inorganic Course, questions like this were answered by linking $\Delta G_m^\ominus$ with a more familiar measure of chemical equilibrium – the **equilibrium constant**, K. At that stage, the formal relation between them was written as follows:

$$\Delta G_m^\ominus = -RT \ln K \tag{9}$$

where R (= 8.314 J K^{-1} mol^{-1}) is the gas constant.

Notice that here, and throughout this Course, we use the symbol 'ln' to represent the *natural* logarithm, or logarithm to the base e, of any quantity: it is related to the more familiar logarithm to the base ten (denoted 'log') as $\ln x = 2.303 \log x$.

■ Use equation 9 to calculate the equilibrium constant at 298.15 K for reaction 1, given $\Delta G_m^\ominus = -32.8$ kJ mol^{-1}

■ $\ln K = +\dfrac{32.8 \times 10^3 \text{ J mol}^{-1}}{8.314 \text{ J K}^{-1}\text{ mol}^{-1} \times 298.15 \text{ K}}$

$= 13.232$

Taking the 'inverse' natural logarithm (on your calculator), $K = 5.58 \times 10^5$.*

---

* To help you check your calculations, we shall sometimes (as here) quote the result of an intermediate step. Any such result will be quoted to a sufficient number of figures to ensure that the *final* answer is correct (within the accuracy of the original data). Make sure you do the same! In other words, never 'round-off' an intermediate result – only the final answer. In this case, for example, the 'rounded' figure ln $K = 13.2$, say, gives $K = 5.40 \times 10^5$.

Table 3 Some numerical values of $\Delta G_m^\ominus$ with corresponding values of $\ln K$ and K at 298.15 K (from equation 9 with $R = 8.314\,\text{J K}^{-1}\,\text{mol}^{-1}$).

$\Delta G_m^\ominus /\text{kJ mol}^{-1}$	$\ln K$	K
−500	201.7	4.0×10^{87}
−100	40.34	3.3×10^{17}
−50	20.17	5.8×10^{8}
−10	4.034	5.6×10
−5	2.017	7.5
0	0	1
5	−2.017	1.3×10^{-1}
10	−4.034	1.8×10^{-2}
50	−20.17	1.7×10^{-9}
100	−40.34	3.0×10^{-18}
500	−201.7	2.5×10^{-88}

This result, taken together with the more extensive compilation in Table 3, serves to re-emphasize that $\Delta G_m^\ominus$ is really just a measure of the equilibrium constant for a reaction at a particular temperature: the more negative is $\Delta G_m^\ominus$, the larger is K. Equally, a moderately positive value of $\Delta G_m^\ominus$ (up to +10 kJ mol⁻¹, say, at 298.15 K) simply implies that K is small, not that the reaction is completely impossible.

As you will see shortly, equation 9 is, in fact, the key to a more quantitative treatment of chemical equilibrium – an analysis that allows us to *predict* the equilibrium yield of product from a reaction. Before embarking on that analysis, however, we are now in a position to establish a very important general result.

4 THE TEMPERATURE DEPENDENCE OF THE EQUILIBRIUM CONSTANT

Notice that our two fundamental relations (equations 2 and 9) can be combined to yield a third expression, as follows:

$$-RT \ln K = \Delta H_m^\ominus - T \Delta S_m^\ominus \tag{10}$$

or, on dividing through by $(-RT)$,

$$\ln K = -\frac{\Delta H_m^\ominus}{RT} + \frac{\Delta S_m^\ominus}{R} \tag{11}$$

Obviously, this provides a sort of 'short-cut' to the calculation of equilibrium constants at 298.15 K from thermodynamic data. But the implications of equation 11 are more wide-ranging than this. They are seen most clearly if we recall a general approximation that was introduced in the Second Level Inorganic Course:

> Provided the physical states of the reactants and products do not change, then it is usually a good approximation to assume that the values of $\Delta H_m^\ominus$ and $\Delta S_m^\ominus$ for a reaction do not change with temperature.

Assuming this to be a reasonable approximation*, it then follows that we can substitute in equation 11 the values of $\Delta H_m^\ominus$ and $\Delta S_m^\ominus$ at 298.15 K, and these values will continue to be appropriate at some other temperature, T. Thus, the equilibrium constant at the other temperature, $K(T)$, is given by

$$\ln K(T) = -\frac{\Delta H_m^\ominus (298.15\ \text{K})}{RT} + \frac{\Delta S_m^\ominus (298.15\ \text{K})}{R} \qquad (12)$$

which can be recast in the following form:

$$\ln K(T) = \left\{ \left(-\frac{\Delta H_m^\ominus (298.15\ \text{K})}{R} \right) \times \left(\frac{1}{T} \right) \right\} + \left\{ \frac{\Delta S_m^\ominus (298.15\ \text{K})}{R} \right\} \qquad (13)$$

This expression can be used to elicit a very important generalization about the temperature dependence of the equilibrium constant for a reaction.

■ Concentrate on the right-hand side of equation 13. Which of the two terms in curly brackets will determine how the equilibrium constant of a given reaction will change as the temperature changes?

▪ The first term. Irrespective of the sign or size of $\Delta S_m^\ominus (298.15\ \text{K})$ for the reaction, the value of the second term is fixed and does not depend on the temperature.

To take the argument a stage further, concentrate now on the first term. Suppose that the temperature T increases: then its reciprocal ($1/T$) must decrease, so that the first term as a whole must become *smaller* in magnitude. But the way this change affects the size of the equilibrium constant itself depends on the *sign* of the enthalpy change, $\Delta H_m^\ominus (298.15\ \text{K})$.

Take the synthesis of ammonia, reaction 1, for example. In this case, $\Delta H_m^\ominus (298.15\ \text{K}) = -92.0\ \text{kJ mol}^{-1}$: the reaction is *exothermic*. This, in turn, means that the first term in equation 13 must always have a *positive* value, but it will become progressively *smaller* as the temperature rises. We conclude, therefore, that the value of $\ln K$ for reaction 1 must become *less positive* as T increases, so the equilibrium constant itself must get smaller.

STUDY COMMENT   The following SAQ should help to convince you about this slightly involved argument. Part (b) also requires you to interpret correctly the label attached to the axis of a graph. If you have not already done so, it would be a good plan to consult Section 2 of the AV Booklet, and listen to the accompanying tape sequence (band 2 on audiocassette 1).

SAQ 3 (a)   For reaction 1, $\Delta H_m^\ominus (298.15\ \text{K}) = -92.0\ \text{kJ mol}^{-1}$ and $\Delta S_m^\ominus = -198.7\ \text{J K}^{-1}\ \text{mol}^{-1}$. Use this information to calculate $\ln K$ and hence K at $T = 1\,000\ \text{K}$. Compare your answer with the value $K(298.15\ \text{K}) = 5.58 \times 10^5$ that you calculated in Section 3.

(b)   The corresponding value of $\ln K$ at 500 K is plotted against reciprocal temperature in Figure 1 (*overleaf*). Enter your value from part (a). (*Take careful note of the label on the horizontal axis.*) Would you be justified in joining these two points with a straight line? Explain your answer. How would you expect a plot of $\ln K$ against $1/T$ for an *endothermic* reaction to differ from your completed version of Figure 1?

---

* Strictly speaking, this approximation holds only over a limited temperature range for most reactions, say 100 K or so. However, it often *apparently* holds over much wider ranges, because in many cases the actual changes in $\Delta H_m^\ominus$ and $\Delta S_m^\ominus$ with change of temperature very nearly cancel each other out when used in equation 11.

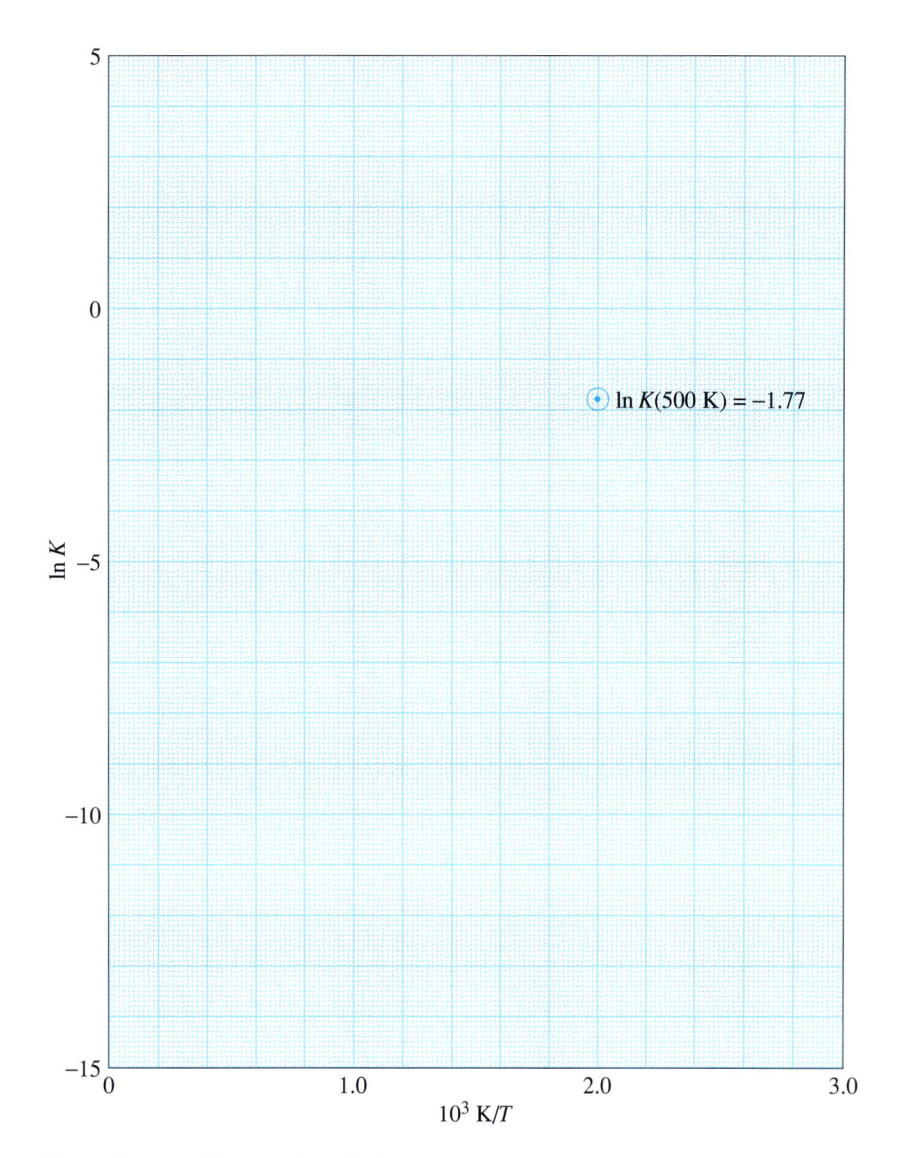

Figure 1 Plot of ln K against $1/T$ for reaction 1, showing the value of ln K at $T = 500$ K.

Notice that the phrase 'less positive' also covers the situation when ln K actually has a negative value at some particular temperature, and so becomes progressively 'more negative' as the temperature rises: this should be clear from your completed version of Figure 1.

Can you now see the broader implications in equation 12 (or 13)?

There are two related points. First, the conclusion reached above is perfectly general: the way the equilibrium constant of *any* reaction changes with temperature depends *solely* on the *sign* of $\Delta H_m^\ominus$ for that reaction.

> When the temperature is raised, K decreases for *any* exothermic reaction ($\Delta H_m^\ominus < 0$), but increases for any endothermic reaction ($\Delta H_m^\ominus > 0$).

Second, equation 12 provides a *quantitative* (albeit approximate) route to the calculation of equilibrium constants at temperatures *other than* 298.15 K. We shall make good use of these values, and of plots like Figure 1 derived from them, in later Sections.

SAQ 4   Consider again reaction 6 in SAQ 1:

$$CH_3OH(g) + \tfrac{1}{2}O_2(g) = HCHO(g) + H_2O(g) \qquad (6)$$

with $\Delta H_m^\ominus(298.15\,\text{K}) = -149.7\,\text{kJ mol}^{-1}$ and $\Delta S_m^\ominus = 65.3\,\text{J K}^{-1}\,\text{mol}^{-1}$.

Use this information to predict the effect of increasing temperature on (a) the equilibrium constant, and (b) the value of $\Delta G_m^\ominus$, for this reaction. (Hint: in tackling part (b), concentrate on the expression in equation 2.)

At first sight, the answers to parts (a) and (b) of SAQ 4 may seem to be in conflict. According to our original criterion (simply the size and sign of $\Delta G_m^\ominus$), reaction 6 could *appear* to become more favourable as the temperature rises, yet its equilibrium constant decreases. The key to this apparent dilemma lies with the relation $\Delta G_m^\ominus = -RT\ln K$ (equation 9) or $\ln K = -\Delta G_m^\ominus/RT$. The fact that T features explicitly in this expression means that even if $\Delta G_m^\ominus$ becomes more negative as T increases, the value of $-\Delta G_m^\ominus/RT$ may still become less positive and so K (the *true* measure of equilibrium at *any* temperature) falls. (If need be, convince yourself of this by working out one or two values of $\Delta G_m^\ominus(T)$ and $K(T)$ for reaction 6.)

In conclusion: although the value of $\Delta G_m^\ominus$ at a particular temperature is certainly a measure of the feasibility of a reaction *at that temperature*, it can be very misleading to draw conclusions from a comparison between values of $\Delta G_m^\ominus$ at *different* temperatures. Always remember that it is the variation of K, as determined by the sign of $\Delta H_m^\ominus$, that actually defines whether a reaction will become more or less favourable with changing temperature.

5 THE EQUILIBRIUM CONSTANT
AND THE EQUILIBRIUM YIELD

To summarize our conclusions so far: provided only that the requisite data are available, the relations of thermodynamics allow us to calculate the equilibrium constant for any reaction, at (effectively) any temperature. This obviously represents an important step toward being able to predict the equilibrium yield for a reaction under specified conditions. But to take the analysis a stage further, we must first revise our ideas about the equilibrium constant: how does the value of K characterize the actual *composition* of an equilibrium mixture?

STUDY COMMENT   Before reading further, do SAQ 5 which revises what you already know about this topic from the prerequisite courses.

SAQ 5 (revision)   Write expressions for the equilibrium constant for each of the following reactions, in terms of the concentrations of reactants and products:

(a) $HF(aq) = H^+(aq) + F^-(aq)$ \hfill (14)

(b) $N_2(g) + 3H_2(g) = 2NH_3(g)$ \hfill (1)

(c) $\tfrac{1}{2}N_2(g) + \tfrac{3}{2}H_2(g) = NH_3(g)$ \hfill (3)

5.1 The equilibrium constant for a gas reaction

For reactions in solution, such as the dissociation of hydrofluoric acid (equation 14), the most direct experimental measure of composition is undoubtedly the molar concentration. Equilibrium constants expressed in these terms are usually denoted by attaching a subscript c to the symbol K, as K_c.* But here we are mainly concerned with reactions that take place in the gas phase. For such reactions, the composition of the equilibrium mixture is usually described in terms of an alternative, and more convenient, quantity – the **partial pressure**. The partial pressure can be defined as follows. Suppose you have a mixture of two gases – call them A and B, say – containing an amount (usually expressed in moles) n_A of A and an amount n_B of B. Suppose further that each gas obeys the **ideal gas equation**. The ideal gas equation is a sort of 'distillation' of countless measurements of the properties of a large number of gases. Strictly speaking, it represents the behaviour of a hypothetical ideal or *perfect* gas, in which there are *no* interactions between the atoms or molecules that make up the gas. Experiments show that *real* gases behave increasingly like an ideal gas as the pressure is lowered, and that the following equation is often a reasonable approximation under ambient conditions:

$$pV = nRT \qquad \text{ideal gas equation} \qquad (15)$$

where p is the pressure, V the volume and n the amount of gas. Then, the partial pressures, $p(A)$ and $p(B)$, of A and B are defined by the relations:

$$p(A) = \frac{n_A RT}{V}; \quad p(B) = \frac{n_B RT}{V} \qquad (16)$$

where V is the volume of the mixture. Moreover, if the *mixture* also obeys equation 15, the total pressure, p_{tot}, is simply the *sum* of the partial pressures: in this case

$$p_{tot} = p(A) + p(B) \qquad (17)$$

■ Convince yourself that this is so.

▪ From equation 15, $p_{tot} = n_{tot}RT/V$. But $n_{tot} = n_A + n_B$, so

$p_{tot} = (n_A + n_B)RT/V$

$\quad = (n_A RT/V) + (n_B RT/V)$

$\quad = p(A) + p(B)$

Evidently, the *partial* pressure is fulfilling the role its name suggests! It is that part of the total pressure 'contributed', as it were, by the substance in question. Equation 17 is just a special case of a completely general relation, known as **Dalton's law of partial pressures**. We shall make good use of it in a moment.

To return to the problem of equilibrium constants: according to the definitions in equation 16, at a given temperature the partial pressure of a gas is directly proportional to the amount of that gas present in a mixture – hence its use to describe the composition of a gas mixture, and to define a 'new' equilibrium constant, denoted K_p, for gas reactions. As before, the expression for K_p follows directly from the stoichiometry of the balanced reaction equation, only this time partial pressures are used instead of concentrations.†

■ Using the notation $p(N_2)$, $p(H_2)$, etc., write down an expression for K_p for our example reaction, equation 1:

$$N_2(g) + 3H_2(g) = 2NH_3(g) \qquad (1)$$

---

* You may be a little puzzled at this point. Values of K calculated from thermodynamic data have hitherto been quoted as pure numbers, with no units. Yet you no doubt recall that 'experimental' equilibrium constants, like K_c for example, generally do have units associated with them. This is a very important question, which is taken up in detail in Section 5.2.4.

† From equation 16, it should be clear that an equilibrium constant expressed in terms of partial pressures can be related to one expressed in terms of concentrations c ($c_A = n_A/V$, etc.).

■ The expression is

$$K_p = \frac{\{p(NH_3)\}^2}{p(N_2)\{p(H_2)\}^3} \tag{18}$$

5.2 Calculating the equilibrium yield

So far so good. But our aim is to calculate the equilibrium yield of product (ammonia, in this case) under a range of reaction conditions. And to do that, we need to resolve a number of issues. First, we need a more precise definition of equilibrium yield. Second, we need to relate this definition in some way to the expression in equation 18. Third, since our aim is to *predict* equilibrium yields, we need access to values of K_p. Given the discussion in previous Sections, it should come as no surprise that this, in turn, revolves around the connection between K_p, on the one hand, and the equilibrium constant derived from thermodynamic data (hitherto denoted as K), on the other.

5.2.1 The equilibrium yield

To deal with these issues in turn: first, it must be said that the yield from a reaction can be defined in a number of different ways. Here, we shall adopt the following definition of the **equilibrium yield** of the desired product *in a gaseous reaction*, partly because it simplifies quite considerably the subsequent analysis:

$$\text{equilibrium yield of product} = \frac{p(\text{product})}{p_{\text{tot}}} \tag{19}$$

where p_{tot} is now the total pressure of the equilibrium mixture.

At first sight this definition may seem a trifle bizarre! It becomes less so once we recall that pressures are proportional to amounts. Thus, the ratio in equation 19 effectively expresses the amount of product in the equilibrium mixture as a *fraction* of the total amount of the different gases then present: *we shall give it the symbol y.*

■ According to our definition, what are the *theoretical* limits to the value of y for the synthesis of ammonia (equation 1)?

■ y is a dimensionless fraction. In this case ammonia is the only product, so the value of y must lie between zero (no product) and one (all product).

But notice that this will not always be so. If the reaction in question leads to *several* products, then even complete reaction will produce a *mixture* of substances. The maximum value of y for the product of interest will then be *less* than one. (This should become clearer when you work through the example in SAQ 6 at the end of the following subsection.)

5.2.2 The expression for K_p

The second point noted above is a little more involved. Look back at equation 18. Assuming that values of K_p are available, our aim is to use this expression to determine the equilibrium yield, y. But to do so, we must first rewrite the right-hand side in terms of y. The trick lies with Dalton's law of partial pressures.

Consider the following imaginary experiment: nitrogen and hydrogen are mixed in *stoichiometric proportions* (that is, in the ratio 1 : 3), and the reaction in equation 1 is allowed to come to equilibrium. Suppose further that the temperature and overall pressure are kept constant throughout.*

■ Write an expression for $p(NH_3)$ at equilibrium in terms of y.

---

* The thermodynamic analysis that follows is valid, notwithstanding any *practical* difficulties that may bedevil attempts to do this imaginary experiment in practice.

- This follows directly from the general definition in equation 19. In this case, $y = p(NH_3)/p_{tot}$, so

$$p(NH_3) = yp_{tot} \tag{20}$$

- Now write an expression for the total pressure of the equilibrium mixture, in terms of $p(H_2)$, etc.

- From Dalton's law

$$p_{tot} = p(N_2) + p(H_2) + p(NH_3) \tag{21}$$

Finally, see if you can work out simple expressions for $p(N_2)$ and $p(H_2)$ in terms of p_{tot} and y. Remember that N_2 and H_2 were mixed in the ratio 1 : 3.

The first step is to 'isolate' the reactants in equation 21,

$$p(N_2) + p(H_2) = p_{tot} - p(NH_3)$$
$$= p_{tot} - yp_{tot}$$
$$= p_{tot}(1 - y) \tag{22}$$

where we have substituted for $p(NH_3)$ from equation 20.

Our hint holds the clue to the next step. *If the reactants are mixed initially in stoichiometric proportions, then $p(H_2) = 3p(N_2)$ throughout the reaction.* Substituting in equation 22, it follows that at equilibrium,

$$p(N_2) + 3p(N_2) = 4p(N_2) = p_{tot}(1 - y)$$

So

$$p(N_2) = \frac{p_{tot}(1-y)}{4}; \; p(H_2) = \frac{3p_{tot}(1-y)}{4} \tag{23}$$

The final step is to substitute the expressions in equations 20 and 23 back into equation 18, to give:

$$K_p = \frac{(yp_{tot})^2}{\left\{\frac{p_{tot}(1-y)}{4}\right\}\left\{\frac{3p_{tot}(1-y)}{4}\right\}^3} = \frac{y^2 p_{tot}^2}{\frac{27}{256}(1-y)^4 p_{tot}^4}$$

So

$$K_p = \left\{\frac{256 y^2}{27(1-y)^4}\right\} \times \frac{1}{p_{tot}^2} \tag{24}$$

Do not be too alarmed by the complexity of this expression, and the problem of 'solving for y' that it apparently poses. You will see later that there is no need to tackle the calculation of y in this 'head-on' fashion. However, our final expression does serve to highlight a very important point. Notice that the overall pressure p_{tot} appears explicitly in equation 24.

Can you see what this implies as far as the value of y is concerned?

Like all equilibrium constants, the value of K_p depends only on the temperature*. The form of equation 24 then implies that the actual equilibrium yield, at any given temperature, will depend on the overall pressure at which the reaction takes place.

By referring to equation 24, decide whether the value of y will increase or decrease with increasing pressure (at constant T).

Concentrate on the right-hand side of equation 24. If the pressure increases, then $(1/p_{tot}^2)$ will decrease. For K_p to remain constant (as it must), the term in curly brackets must *increase*, and the only way this can happen is if y itself increases.

---

* In fact, this statement is strictly true only if the gases involved in the reaction behave ideally. We shall assume this to be the case, but we do come back to this point later on in this Section.

Specifically, if y increases, then y^2 will increase, but $(1 - y)$ and hence $(1 - y)^4$ will decrease: both changes increase the term in curly brackets, as required. (You may like to convince yourself of this by taking one or two values of y, in the range $0 < y < 1$, and using your calculator to see what happens to the term in curly brackets.)

■ Does this conclusion accord with the application of Le Chatelier's principle to the equilibrium in equation 1?

▪ Yes. According to Le Chatelier's principle, the influence of an *increase* in pressure will be lessened by a reduction in the volume. The equilibrium will shift to the side having the smaller number of gaseous molecules – the right-hand side (ammonia) in this case.

This should not be too surprising. If you follow carefully the steps whereby we arrived at equation 24, you will see that the way p_{tot} features in this expression is a direct consequence of the stoichiometry of the balanced reaction equation. Indeed, we should stress that the procedure outlined above provides a completely general route to the expression for K_p for *any* gas reaction: *the only proviso is that the reactants are mixed initially in stoichiometric proportions* (at constant temperature and pressure, of course).

STUDY COMMENT The following SAQ gives you a chance to practise this technique for yourself. Do not miss it out!

SAQ 6 Vinyl chloride (chloroethene), **2**, is produced on an industrial scale by the gas-phase pyrolysis of 1,2-dichloroethane, **1**, as follows:

$$CH_2Cl-CH_2Cl(g) = CH_2=CHCl(g) + HCl(g) \qquad (25)$$
$$ \mathbf{1} \mathbf{2}$$

where $\Delta H_m^\ominus (298.15\ K) = +72.5\ kJ\ mol^{-1}$.

Use the procedure outlined above to develop an expression for K_p in terms of the equilibrium yield y of vinyl chloride, and the overall pressure p_{tot}. What is the maximum theoretical value of y in this case?

Would you expect y to increase or decrease with: (a) increasing pressure; (b) increasing temperature?

5.2.3 The effect of pressure on the equilibrium yield

To generalize the link between the stoichiometry of a reaction and the way p_{tot} features in the expression for K_p, notice that in both of the examples considered above (reactions 1 and 25), K_p has the form

$$K_p = (\text{function of } y) \times (p_{tot})^\Sigma \qquad (26)$$

where the index Σ (Greek sigma) comes from the stoichiometry of the balanced reaction equation, *as written*. Thus, for a general reaction,

$$a\text{A} + b\text{B} + \ldots = p\text{P} + q\text{Q} + \ldots \qquad (4)$$

$$\Sigma = (p + q + \ldots) - (a + b + \ldots) \qquad (27)$$

Check that this gives the same result as the expressions in equation 24 and in your answer to SAQ 6.

To summarize: our more detailed treatment of chemical equilibrium has enabled us to rationalize qualitative insights derived from Le Chatelier's principle.

> When the pressure is raised, the equilibrium yield of product increases for any reaction with $\Sigma < 0$ (for example, the synthesis of ammonia), but decreases for any reaction with $\Sigma > 0$ (see the example in SAQ 6).

5.2.4 The standard equilibrium constant

As far as the synthesis of ammonia is concerned, you should now be clear about *why* the equilibrium yield is reduced by raising the temperature, but increased by raising the pressure. But to *quantify* the effects of temperature and pressure on this (or any other) gaseous equilibrium requires values of K_p. This brings us back to the third point in our introductory remarks – the connection between an equilibrium constant defined in terms of partial pressures (K_p), and one effectively defined in terms of thermodynamic quantities, via the relation:

$$\ln K(T) = -\frac{\Delta H_m^\ominus}{RT} + \frac{\Delta S_m^\ominus}{R} \tag{11}$$

This raises a slight problem.

■ Look back at equation 24. What are the dimensions of K_p as given by this expression?

▪ Since y is defined to be a dimensionless fraction (equation 19), K_p has the dimensions (pressure)$^{-2}$ in this case.

In view of the discussion above (and the general expression in equation 26), it should be clear that the dimensions of K_p will *always* depend on the stoichiometry of the reaction equation.

■ But what are the dimensions of the two terms on the right-hand side of equation 11?

▪ $\Delta H_m^\ominus$ has dimensions of molar energy, as does RT; both $\Delta S_m^\ominus$ and R have dimensions of molar energy per degree. Thus both terms are dimensionless, as indeed they must be. The natural logarithm *and its inverse* (hence the value of K) must both be pure numbers. (See also Table 3 and your answer to SAQ 3.)

This, then, is the crux of the problem: it can be resolved very simply as follows. First, the dimensionless equilibrium constant defined by equation 11 is formally called the **standard** (or **thermodynamic**) **equilibrium constant**, and is represented by the symbol $K^\ominus$. Thus equation 11 (and the alternative forms of it in Section 4) should strictly speaking be written:

$$\ln K^\ominus(T) = -\frac{\Delta H_m^\ominus}{RT} + \frac{\Delta S_m^\ominus}{R} \tag{28}$$

It follows that equilibrium constants calculated from thermodynamic data are actually values of $K^\ominus$ at the temperature in question. For our reaction (equation 1) for instance,

$$K^\ominus(298.15 \text{ K}) = 5.58 \times 10^5$$

The link between $K^\ominus$ and the corresponding value of K_p rests on a very important convention:

> $K^\ominus$ is related to K_p by dividing each partial pressure in the expression for K_p by a standard value of pressure $p^\ominus$.

By international agreement, the standard pressure is taken to be **one bar** (1 bar) which is defined to be exactly 10^5 Pa (= 100 kPa). This is only slightly less than the atmosphere (1 atm = 101.325 kPa = 1.013 25 bar) – an important practical unit of pressure. The bar has the great advantage of being immediately converted into fundamental SI units (1 Pa = 1 kg m^{-1} s^{-2}), without the nuisance of a lengthy numerical factor, whence its choice as the standard pressure.

Now look back at the general expression for K_p in equation 26. The reasoning that led to this expression suggests that the force of the convention above is to define $K^\ominus$ as follows:

$$K^\ominus = (\text{function of } y) \times \left(\frac{p_{\text{tot}}}{p^\ominus}\right)^\Sigma$$

or, from equation 26,

$$K^\ominus = \frac{K_p}{(p^\ominus)^\Sigma} \qquad (29)$$

where, as before, Σ is determined by the stoichiometry of the reaction equation (equation 27).

■ According to this convention, what is the value of K_p for reaction 1 at 298.15 K?

▪ From equation 29, $K_p = K^\ominus \times (p^\ominus)^\Sigma$, and $\Sigma = -2$ for reaction 1, so

$K_p(298.15\text{ K}) = 5.58 \times 10^5 \times (1\text{ bar})^{-2}$

$\qquad = 5.58 \times 10^5 \text{ bar}^{-2}$

$\qquad = 5.58 \times 10^5 \times (10^5 \text{ Pa})^{-2}$

$\qquad = 5.58 \times 10^{-5} \text{ Pa}^{-2}$ (a very different value!)

Notice that this result suggests a rather simpler interpretation of the convention above: since $p^\ominus$ is chosen to be *one bar*, then $K^\ominus$ is just the numerical magnitude of K_p, *provided pressures are measured in bars*. As implied by equation 29, the units of K_p are then just (bar)$^\Sigma$.

Strictly speaking, there is rather more to this 'convention' than meets the eye. In writing an expression for K_p in terms of partial pressures, and in then identifying $K^\ominus$ with the magnitude of K_p, we implicitly *assume* that the reaction mixture behaves ideally. We come back to this point in the next Section, and again (in more detail) in Block 7. There, too, you will meet (and subject to critical examination) an analogous convention linking K_c to $K^\ominus$ for solution reactions.

STUDY COMMENT To gain some familiarity with the definition of $K^\ominus$, try the following SAQ.

SAQ 7 What is the value of K_p at 298.15 K for each of the following reactions? Express your answers in both bar and Pa.

(a) $CH_3OH(g) + \tfrac{1}{2}O_2(g) = HCHO(g) + H_2O(g)$; $K^\ominus = 4.41 \times 10^{29}$ (6)

(b) $2N_2O_5(g) = 4NO_2(g) + O_2(g)$; $K^\ominus = 2.60 \times 10^4$ (7)

(c) $2CO(g) + 3H_2(g) = C_2H_2(g) + 2H_2O(g)$; $K^\ominus = 2.37 \times 10^{-5}$ (8)

6 THE SYNTHESIS OF AMMONIA: THE THERMODYNAMIC 'LIMITS OF OPERATION'

By introducing the connection between K_p and $K^\ominus$, we have finally succeeded in establishing a *quantitative* link between thermodynamic data, on the one hand, and the temperature – and pressure – dependence of equilibrium yields, on the other. Indeed, with the background provided by the previous Sections, you should be able to tackle the problem of predicting equilibrium yields for yourself. As far as the synthesis of ammonia is concerned, the important points are summarized in Box 1. *Make sure you try SAQ 8 before moving on.*

SAQ 8 Suppose that you want to establish the conditions under which it is possible to achieve a 20% equilibrium yield of ammonia, that is $y = 0.2$, from reaction 1.

(a) Assuming an overall pressure of 1 bar, and that the reactants are initially present in stoichiometric proportions, determine the value of K_p, and hence $K^\ominus$, necessary to ensure this yield. Then use your completed version of Figure 1 (Section 4) to determine the temperature at which $K^\ominus$ has this value.

(b) Now repeat the calculation in part (a), but for an overall pressure of 10 bar.

Box 1 The synthesis of ammonia

Balanced reaction equation:

$$N_2(g) + 3H_2(g) = 2NH_3(g) \tag{1}$$

Thermodynamic properties of reaction 1:

$\Delta H_m^\ominus (298.15 \text{ K}) = -92.0 \text{ kJ mol}^{-1}$

$\Delta S_m^\ominus (298.15 \text{ K}) = -198.7 \text{ J K}^{-1} \text{ mol}^{-1}$

$\Delta G_m^\ominus (298.15 \text{ K}) = -32.8 \text{ kJ mol}^{-1}$

$K_p = K^\ominus \times (p^\ominus)^\Sigma$; $p^\ominus = 1$ bar; $\Sigma = -2$

Expression for K_p:

$$K_p = \left\{ \frac{256 y^2}{27(1-y)^4} \right\} \times \frac{1}{p_{tot}^2} \tag{24}$$

where y is the equilibrium yield of NH_3 from a stoichiometric mixture, at a constant overall pressure, p_{tot}.

The procedure you followed in answering SAQ 8 represents the 'indirect' approach to calculating equilibrium yields referred to earlier. This technique is completely general: the temperature necessary to ensure a specified equilibrium yield, at a given total pressure, can always be 'read' from a plot of $\ln K^\ominus$ against $1/T$, like your completed version of Figure 1. But there is a simpler way. *If $\Delta H_m^\ominus$ and $\Delta S_m^\ominus$ are assumed to be independent of temperature* (as here), then the expression in equation 28 allows such temperatures to be determined, without recourse to a graphical treatment.

■ If $K^\ominus$ is the required value of the equilibrium constant, use equation 28 to obtain a formula for the corresponding temperature.

$$\ln K^\ominus = -\frac{\Delta H_m^\ominus}{RT} + \frac{\Delta S_m^\ominus}{R} \tag{28}$$

- Multiplying through equation 28 by RT gives

$$RT \ln K^\ominus = -\Delta H_m^\ominus + T\Delta S_m^\ominus$$

So, collecting the terms containing T on the right-hand side, and moving $\Delta H_m^\ominus$ over to the left,

$$\Delta H_m^\ominus = T\Delta S_m^\ominus - RT \ln K^\ominus = T(\Delta S_m^\ominus - R \ln K^\ominus)$$

$$\text{or } T = \frac{\Delta H_m^\ominus}{\Delta S_m^\ominus - R \ln K^\ominus} \tag{30}$$

(Check that equation 30 does indeed give values of T that agree with your estimates in the answer to SAQ 8.)

At this stage, it should be clear that equilibrium yields of ammonia can be computed over predetermined pressure and derived temperature ranges: the resulting values constitute the **thermodynamic 'limits of operation'** referred to in Section 1. To examine this idea further, the results of our (more extensive) calculations are collected in Figure 2. Here, the full curves were derived in the same way as your answers to SAQ 8: the thermodynamic data used are approximate in two senses.

First, $\Delta H_m^\ominus$ and $\Delta S_m^\ominus$ for reaction 1 are assumed constant throughout the temperature range from 298.15 K to 900 K: this approximation becomes increasingly unreliable as the temperature increases. The second point is one we touched on earlier – namely the assumption of ideal behaviour. In fact, this assumption breaks down seriously at the very high pressures recorded in Figure 2, as a comparison with the experimental curves (broken lines) clearly demonstrates. Notice, however, that the two sets of data do coincide at low pressure and at moderate temperatures. This provides support for our use of approximate thermodynamic data – and for the assumption of ideal behaviour – provided the reaction conditions are not too extreme.

- Are the general features of Figure 2 in accord with the expected behaviour of this equilibrium?

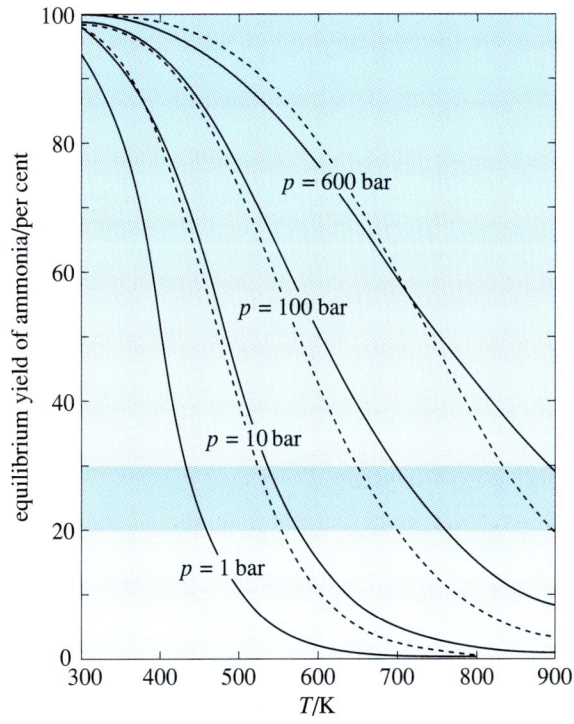

Figure 2 Equilibrium yield of ammonia as a function of temperature and pressure, from an initial mixture containing hydrogen and nitrogen in the ratio 3:1. The full curves were calculated from thermodynamic data (see also the text): the dashed curves are experimental results. The coloured band is explained in the text.

■ Yes. At any given pressure, the equilibrium yield falls off with increasing temperature: indeed, the effect is quite dramatic in this case. Equally, at constant temperature (any vertical line on the Figure), the yield is clearly improved by raising the pressure.

But what about the broader issues raised in our introductory remarks in Section 1? To be more specific: since the first commercial operation of the Haber–Bosch process in 1913, plants have traditionally been designed to run at a temperature in the range 400–540 °C (673–813 K) and a pressure in the range 150–600 bar, in the presence of a solid catalyst composed mainly of iron. To what extent does the thermodynamic information summarized in Figure 2 allow us to rationalize this choice of reaction conditions?

Your initial response to this question may well be rather scathing! After all, it is clear from Figure 2 that the highest equilibrium yields of ammonia are achieved at the lowest temperatures: thus, thermodynamics argues for a low synthesis temperature. This immediately brings us up against the fundamental limitation of *any* purely thermodynamic analysis: it says nothing about the actual time-scale of a reaction. In this case, it contains no hint of the experimental fact that the direct reaction between gaseous nitrogen and hydrogen is immeasurably slow under ambient conditions.

The important general point (stressed in the Second Level Inorganic Course) is that the thermodynamics of a reaction depends only on the initial and final states of the system: thermodynamics tells us nothing about the processes that occur as the reactants are actually *converted* into the products. These processes – the underlying **mechanism** of the reaction – are the province of chemical kinetics. In this case, the mechanism is the process whereby the very strong N≡N and H—H bonds in the N_2 and H_2 molecules are broken, and N—H bonds are formed. Somewhere in this mechanism lies the reason for the 'kinetic stability' of a mixture of nitrogen and hydrogen at normal temperatures.

By the time you have completed Block 5 of this Course, you should have a fair idea of the principles behind two common methods of *speeding up* a reaction like this: raising the temperature; and developing a suitable **catalyst**. Both of these 'kinetic factors' are crucial to the commercial success of the Haber–Bosch process, but especially the latter. Put simply, the *surface* of the iron catalyst effectively acts as a 'meeting place' for the reactants: its presence thus alters the mechanism of the reaction, allowing the necessary changes in bonding to take place more easily, and thereby providing a *faster* reaction pathway. This is the essence of catalysis.

Nevertheless, we should not be too quick to dismiss the hard-won fruits of our thermodynamic analysis. For example, the original catalyst for ammonia synthesis was developed in 1910, three years before the German company, Badische Anilin und Soda Fabrik (BASF), brought the first industrial ammonia plant on stream. A research team (led by Carl Bosch) found that the iron catalyst was substantially improved if small amounts of potassium and aluminium oxides were present. Similar catalysts are still widely used today, although small amounts of other oxides (such as calcium oxide) are usually added as well. Even with such catalysts, however, the operating temperature for ammonia synthesis must still be around 750 K (over 450 °C) in order to achieve an acceptable rate of reaction. It is clear from Figure 2 that there would be little point in running the process at atmospheric pressure, nor even at 10 bar: conditions under which the *equilibrium* yield is less than 5% at 750 K! Rather, the coloured band across Figure 2 indicates that the range of conditions currently used in most plants corresponds to an equilibrium yield of around 20–30%. Higher yields, although desirable, would clearly involve difficult and expensive operating conditions, such as the use of very high pressures.

On the other hand, it is equally clear that if a more 'active' catalyst could be found – and here we take this to mean one that would permit a *lower* operating temperature – then *thermodynamics* would also allow a lower overall pressure, with consequent savings in expensive equipment, energy, etc., and in the maintenance of plant. Recent years have seen some moves in this direction. For example, in the mid 1980s, ICI announced details of a new energy-saving ammonia process (known as the AMV process), following development of a catalyst specifically formulated to give maximum activity at somewhat lower temperatures – in the range 350–430 °C (623–703 K). This, in turn, allows the optimum synthesis pressure to be as low as 70 bar.

It is in this sense that a careful study of the thermodynamics of a reaction can provide guidelines for the industrial chemist – a set of fairly precise limiting conditions within which a given process must operate. Provided the substances involved are reasonably well known (as here), this analysis has the additional advantage of drawing largely on published data, thus reducing the need for expensive and time-consuming experimentation.

Sooner or later, however, such experimentation *must* be undertaken, in order to provide complementary information about the kinetics of the reaction of interest. This is the question – the study of chemical kinetics – that we turn to in the next Block.

Summary of Block 1

1 *If the values of $\Delta H_m^\ominus$ and $\Delta S_m^\ominus$ for a reaction are assumed to be independent of temperature*, then the thermodynamic relations collected in Box 2 provide a route to values of the standard or thermodynamic equilibrium constant, $K^\ominus$, for that reaction at *any* temperature T.

Box 2

$$\Delta G_m^\ominus(T) = \Delta H_m^\ominus(298.15\text{ K}) - T\Delta S_m^\ominus(298.15\text{ K})$$

$$\Delta G_m^\ominus(T) = -RT \ln K^\ominus(T)$$

$$\ln K^\ominus(T) = -\frac{\Delta H_m^\ominus(298.15\text{ K})}{RT} + \frac{\Delta S_m^\ominus(298.15\text{ K})}{R}$$

2 From the expressions in Box 2, when the temperature is raised then $K^\ominus$ decreases for any exothermic reaction ($\Delta H_m^\ominus < 0$) but increases for any endothermic reaction ($\Delta H_m^\ominus > 0$).

3 *Assuming ideal behaviour, and reactants mixed in stoichiometric proportions,* the steps summarized in Box 3 can be used to derive an expression for K_p for any *gas-phase* reaction, in terms of the equilibrium yield y of desired product, and the overall pressure, p_{tot}.

Box 3

For the following reaction:

A(g) + B(g) = 2C(g) + D(g)

$$K_p = \frac{\{p(\text{C})\}^2 p(\text{D})}{p(\text{A})p(\text{B})} \text{ and } K^\ominus = \frac{K_p}{(p^\ominus)^\Sigma}$$

where $\Sigma = 2 + 1 - 1 - 1 = +1$, and $p^\ominus = 1$ bar.

Suppose that the desired product is substance D, then define:

$$y = \frac{p(\text{D})}{p_{\text{tot}}}$$

$\left. \begin{array}{l} p(\text{C}) = 2p(\text{D}) \\ p(\text{A}) = p(\text{B}) \end{array} \right\}$ (from the definition of partial pressure)

$p_{\text{tot}} = p(\text{A}) + p(\text{B}) + p(\text{C}) + p(\text{D})$ (from Dalton's law)

whence: $K_p = (\text{function of } y) \times (p_{\text{tot}})^{+1}$

4   From the expressions in Box 3, the effect of pressure on the equilibrium yield from a gaseous reaction depends solely on the stoichiometry of the balanced reaction equation: when the pressure is raised, y increases for any reaction with $\Sigma < 0$ (for example the synthesis of ammonia), but decreases for any reaction with $\Sigma > 0$ (see the examples in SAQ 6 and Box 3).

5   If pressures are expressed in bars, then $K^{\ominus}$ is just the *magnitude* of K_p at the temperature in question, as indicated in Box 3.

6   Taken together, points 1, 3 and 5 allow equilibrium yields of product to be computed over predetermined pressure and derived temperature ranges. The resulting values constitute the thermodynamic 'limits of operation' for a given reaction, and provide valuable guidelines in the design of an efficient industrial process.

STUDY COMMENT   Do not miss out the Exercise below. It provides an opportunity for you to draw together and apply most of the ideas and techniques developed in Block 1. It also serves to highlight a further limitation of the thermodynamic approach – one that we shall come back to later in the Course.

EXERCISE 1   Any information you need to answer this question should be taken from the S342 *Data Book*.

Methanol is a major industrial chemical. The production process is based on the 'hydrogenation' of carbon monoxide (equation 31), and has been in use since the 1920s.

$$CO(g) + 2H_2(g) = CH_3OH(g) \tag{31}$$

(a) Discuss *briefly* the procedures that could be used to increase the yield of methanol from the reaction in equation 31.

(b) Determine the highest temperature at which it is possible to run this process such that the *equilibrium* yield of methanol is at least 30%. (Assume an overall pressure of 1 bar and reactants mixed in stoichiometric proportions.) *State carefully any assumptions involved in your calculations.*

(c) Industrially, the reaction is carried out at about 500 K in the presence of a solid copper-based catalyst. Determine the pressure necessary to achieve a 30% equilibrium yield at 500 K.

(d) Mixtures of carbon monoxide and hydrogen in various proportions are commonly known as 'synthesis gas'. As indicated in Figure 3, synthesis gas is a highly versatile starting point for chemicals manufacture: not only can it be prepared from just about *any* carbon source, but it also contains the ingredients (C, H and O) of a vast number of organic compounds. *In principle at least*, virtually any hydrocarbon can be produced from synthesis gas (yielding CO_2 or, more commonly, H_2O as by product) – and a whole variety of oxygenated products are also possible: a representative selection of the many possibilities is included in Figure 3.

Bearing in mind the information in Figure 3, what criticism would you make of the simple thermodynamic analysis in parts (a) to (c)?

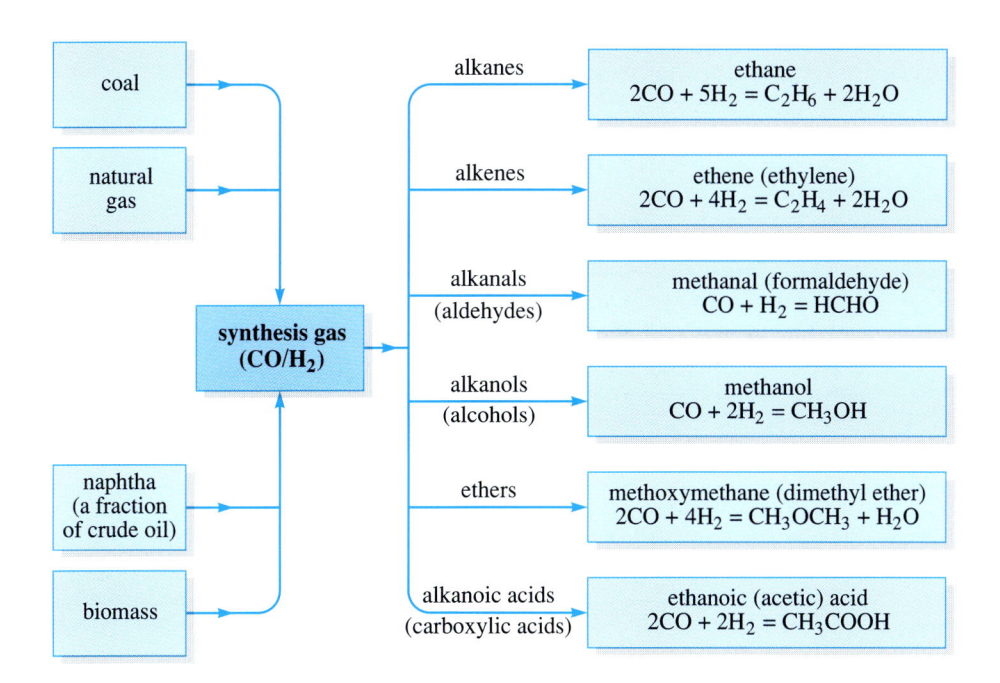

```
coal ──────────┐
               │
natural gas ───┤
               ├──► synthesis gas (CO/H₂) ──┬── alkanes ──────────────► ethane
               │                             │                           $2CO + 5H_2 = C_2H_6 + 2H_2O$
naphtha ───────┤                             │
(a fraction    │                             ├── alkenes ──────────────► ethene (ethylene)
 of crude oil) │                             │                           $2CO + 4H_2 = C_2H_4 + 2H_2O$
               │                             │
biomass ───────┘                             ├── alkanals (aldehydes) ─► methanal (formaldehyde)
                                             │                           $CO + H_2 = HCHO$
                                             │
                                             ├── alkanols (alcohols) ──► methanol
                                             │                           $CO + 2H_2 = CH_3OH$
                                             │
                                             ├── ethers ───────────────► methoxymethane (dimethyl ether)
                                             │                           $2CO + 4H_2 = CH_3OCH_3 + H_2O$
                                             │
                                             └── alkanoic acids ──────► ethanoic (acetic) acid
                                                 (carboxylic acids)      $2CO + 2H_2 = CH_3COOH$
```

Figure 3 A schematic representation of some of the organic compounds that could, *in principle*, be formed from synthesis gas. At present, synthesis gas is usually produced from natural gas or crude oil sources: alternative sources include coal, biomass (any plant material), household garbage – or even manure! (Refer to the glossary of organic compounds in Section 4 of the S342 *Data Book* if you are unfamiliar with any of the chemicals mentioned in the Figure.)

OBJECTIVES FOR BLOCK 1

Now that you have completed Block 1, you should be able to do the following things:

1 Recognize valid definitions of, and use in a correct context, the terms, concepts and principles printed in bold type in the text and collected in the following Table.

List of scientific terms, concepts and principles used in Block 1

Term	Page No.
absolute entropy, $S^\ominus$	8
calculating equilibrium yields from thermodynamic data	20
catalyst	22
criterion for a thermodynamically favourable reaction	6
Dalton's law of partial pressures	14
effect of pressure on equilibrium yield	17
effect of temperature on equilibrium yield	12
equilibrium constant, K	9
–in terms of concentrations, K_c	14
–in terms of partial pressures, K_p	14
equilibrium yield, y	15
Haber–Bosch process	5, 22
ideal gas equation, $pV = nRT$	14
partial pressure	14
relation between $K^\ominus$ and $\Delta G_m^\ominus$	9
relation between $K^\ominus$ and K_p	19
standard (or thermodynamic) equilibrium constant, $K^\ominus$	18
standard enthalpy of formation, $\Delta H_f^\ominus$	7
standard Gibbs function of formation, $\Delta G_f^\ominus$	8
standard molar enthalpy change, $\Delta H_m^\ominus$	7
standard molar entropy change, $\Delta S_m^\ominus$	7
standard molar Gibbs function change, $\Delta G_m^\ominus$	6
temperature dependence of $K^\ominus$	10
thermodynamic data	6
thermodynamic 'limits of operation'	21

2 Predict the effect of changing the temperature on the equilibrium constant for a reaction (given that appropriate enthalpy data are available). (SAQs 3, 4 and 6; Exercise 1)

3 Given a balanced chemical equation for a reaction involving gases, write down an expression for the equilibrium constant K_p in terms of partial pressures and:

(a) use this relation to derive an expression in terms of the equilibrium yield y of product(s), and the overall pressure; (SAQ 6; Exercise 1)

(b) relate the value of K_p at a given temperature to the corresponding value of the standard equilibrium constant $K^\ominus$; (SAQs 7 and 8; Exercise 1)

(c) predict the effect of increasing the pressure on the equilibrium yield of product(s). (SAQ 6; Exercise 1)

BLOCK 1 SCOPE AND LIMITATIONS OF THE THERMODYNAMIC APPROACH

4   For a given gas-phase reaction, select and use appropriate thermodynamic data to:
    (a) calculate the standard equilibrium constant $K^\ominus$, and hence K_p, at any temperature; (SAQ 3; Exercise 1)

    (b) determine appropriate conditions of temperature and pressure to achieve a specified equilibrium yield of a specified product. (SAQ 8; Exercise 1)

5   State the assumptions and approximations involved in the thermodynamic analysis outlined in Objective 4, and indicate briefly the main limitations of this approach. (Exercise 1)

SAQ ANSWERS AND COMMENTS

SAQ 1 (revision of the Second Level Inorganic Course)

(a)

$$\Delta H_m^\ominus = \Delta H_f^\ominus(\text{HCHO, g}) + \Delta H_f^\ominus(\text{H}_2\text{O, g}) - \Delta H_f^\ominus(\text{CH}_3\text{OH, g}) - \tfrac{1}{2}\Delta H_f^\ominus(\text{O}_2, \text{g})$$

$$= \{-108.6 + (-241.8) - (-200.7) - 0\}\,\text{kJ mol}^{-1}$$

$$= -149.7\,\text{kJ mol}^{-1}$$

$$\Delta S_m^\ominus = S^\ominus(\text{HCHO, g}) + S^\ominus(\text{H}_2\text{O, g}) - S^\ominus(\text{CH}_3\text{OH, g}) - \tfrac{1}{2}S^\ominus(\text{O}_2, \text{g})$$

$$= \{218.8 + 188.8 - 239.8 - \tfrac{1}{2}(205.1)\}\,\text{J K}^{-1}\,\text{mol}^{-1}$$

$$= 65.25\,\text{J K}^{-1}\,\text{mol}^{-1}$$

$$\Delta G_m^\ominus = \Delta H_m^\ominus - T\Delta S_m^\ominus$$

$$= (-149.7\,\text{kJ mol}^{-1}) - (298.15\,\text{K}) \times (65.25 \times 10^{-3}\,\text{kJ K}^{-1}\,\text{mol}^{-1})$$

$$= (-149.7 - 19.45)\,\text{kJ mol}^{-1}$$

$$= -169.2\,\text{kJ mol}^{-1}$$

The answers to parts (b) and (c) are calculated in a similar way:

(b)

$$\Delta H_m^\ominus = \{4(33.2) + 0 - 2(11.3)\}\,\text{kJ mol}^{-1} = 110.2\,\text{kJ mol}^{-1}$$

$$\Delta S_m^\ominus = \{4(240.1) + 205.1 - 2(355.7)\}\,\text{J K}^{-1}\,\text{mol}^{-1} = 454.1\,\text{J K}^{-1}\,\text{mol}^{-1}$$

$$\Delta G_m^\ominus = -25.2\,\text{kJ mol}^{-1}$$

(c)

$$\Delta H_m^\ominus = \{226.7 + 2(-241.8) - 2(-110.5) - 3(0)\}\,\text{kJ mol}^{-1} = -35.9\,\text{kJ mol}^{-1}$$

$$\Delta S_m^\ominus = \{200.9 + 2(188.8) - 2(197.7) - 3(130.7)\}\,\text{J K}^{-1}\,\text{mol}^{-1} = -209.0\,\text{J K}^{-1}\,\text{mol}^{-1}$$

$$\Delta G_m^\ominus = +26.4\,\text{kJ mol}^{-1}$$

SAQ 2 (revision of the Second Level Inorganic Course)

Reference to the notes in Section 2 of the S342 *Data Book* should have reminded you that the formation reaction for the aqueous ion Sc^{3+}(aq) is

$$\text{Sc(s)} + 3\text{H}^+(\text{aq}) = \text{Sc}^{3+}(\text{aq}) + \tfrac{3}{2}\text{H}_2(\text{g})$$

Thus $\Delta H_f^\ominus(\text{Sc}^{3+}, \text{aq})$ is the value of $\Delta H_m^\ominus$ for this reaction. It can be calculated from the following equation, which is just a special case of equation 2:

$$\Delta G_f^\ominus = \Delta H_f^\ominus - T\Delta S_f^\ominus$$

or

$$\Delta H_f^\ominus = \Delta G_f^\ominus + T\Delta S_f^\ominus$$

The value of $\Delta G_f^\ominus$ is given in Table 2, but the value of $\Delta S_f^\ominus$ must be *calculated* from the *absolute* entropies of the substances involved in the formation reaction (recall the warning in the main text!). In this case,

$$\Delta S_f^\ominus = S^\ominus(Sc^{3+}, aq) + \tfrac{3}{2}S^\ominus(H_2, g) - S^\ominus(Sc, s) - 3S^\ominus(H^+, aq)$$

$$= \{-255.2 + \tfrac{3}{2}(130.7) - (34.6) - 3(0)\} \, J\,K^{-1}\,mol^{-1}$$

$$= -93.8 \, J\,K^{-1}\,mol^{-1}$$

Then

$$\Delta H_f^\ominus = (-586.6 \, kJ\,mol^{-1}) + (298.15 \, K) \times (-93.8 \times 10^{-3} \, kJ\,K^{-1}\,mol^{-1})$$

$$= -614.6 \, kJ\,mol^{-1}$$

For the bromide, the appropriate formation reaction is

$$Sc(s) + \tfrac{3}{2}Br_2(l) = ScBr_3(s)$$

(Note that $Br_2(l)$ is the reference state of bromine – revealed by the fact that its $\Delta H_f^\ominus$ value is zero in the S342 *Data Book*.)

Thus

$$\Delta S_f^\ominus = S^\ominus(ScBr_3, s) - S^\ominus(Sc, s) - \tfrac{3}{2}S^\ominus(Br_2, l)$$

$$= \{(167.4) - (34.6) - \tfrac{3}{2}(152.2)\} \, J\,K^{-1}\,mol^{-1}$$

$$= -95.5 \, J\,K^{-1}\,mol^{-1}$$

Then

$$\Delta G_f^\ominus = \Delta H_f^\ominus - T\Delta S_f^\ominus$$

$$= (-743.1 \, kJ\,mol^{-1}) - (298.15 \, K) \times (-95.5 \times 10^{-3} \, kJ\,K^{-1}\,mol^{-1})$$

$$= -714.6 \, kJ\,mol^{-1}$$

SAQ 3 (Objectives 2 and 4a)

(a) According to equation 12 (or 13),

$$\ln K \,(1\,000 \, K) = \left\{\left(-\frac{\Delta H_m^\ominus(298.15\,K)}{R}\right) \times \left(\frac{1}{10^3 \, K}\right)\right\} + \left\{\frac{\Delta S_m^\ominus(298.15\,K)}{R}\right\}$$

$$= -\left\{\frac{(-92.0 \times 10^3 \, J\,mol^{-1})}{(8.314 \times 10^3 \, J\,mol^{-1})}\right\} + \left\{\frac{(-198.7 \, J\,K^{-1}\,mol^{-1})}{(8.314 \, J\,K^{-1}\,mol^{-1})}\right\}$$

$$= (+11.065\,7 - 23.899\,4) = -12.833\,7$$

So $K = 2.67 \times 10^{-6}$

Comparing this value with $K(298.15 \, K) = 5.58 \times 10^5$ from Section 3, it is clear that in this case the equilibrium constant does indeed fall with increasing temperature.

(b) The completed graph is shown in Figure 4. As indicated in the question, it is important that you understand the meaning of the label on the horizontal axis: here, it tells you that any point on this axis represents the numerical value of the quantity $(10^3 \, K/T)$. Thus, to plot the point calculated in part (a), you need to work out the corresponding value of this quantity, as:

$$T = 1\,000 \, K$$

So $\dfrac{1}{T}$ or $1/T = \dfrac{1}{10^3 \, K} = 1.0 \times 10^{-3} \, K^{-1}$

Multiplying through by 10^3 gives

$$10^3/T = 1.0 \, K^{-1}$$

or, on multiplying both sides by K,

$$10^3 \, K/T = 1.0$$

as indicated in Figure 4.

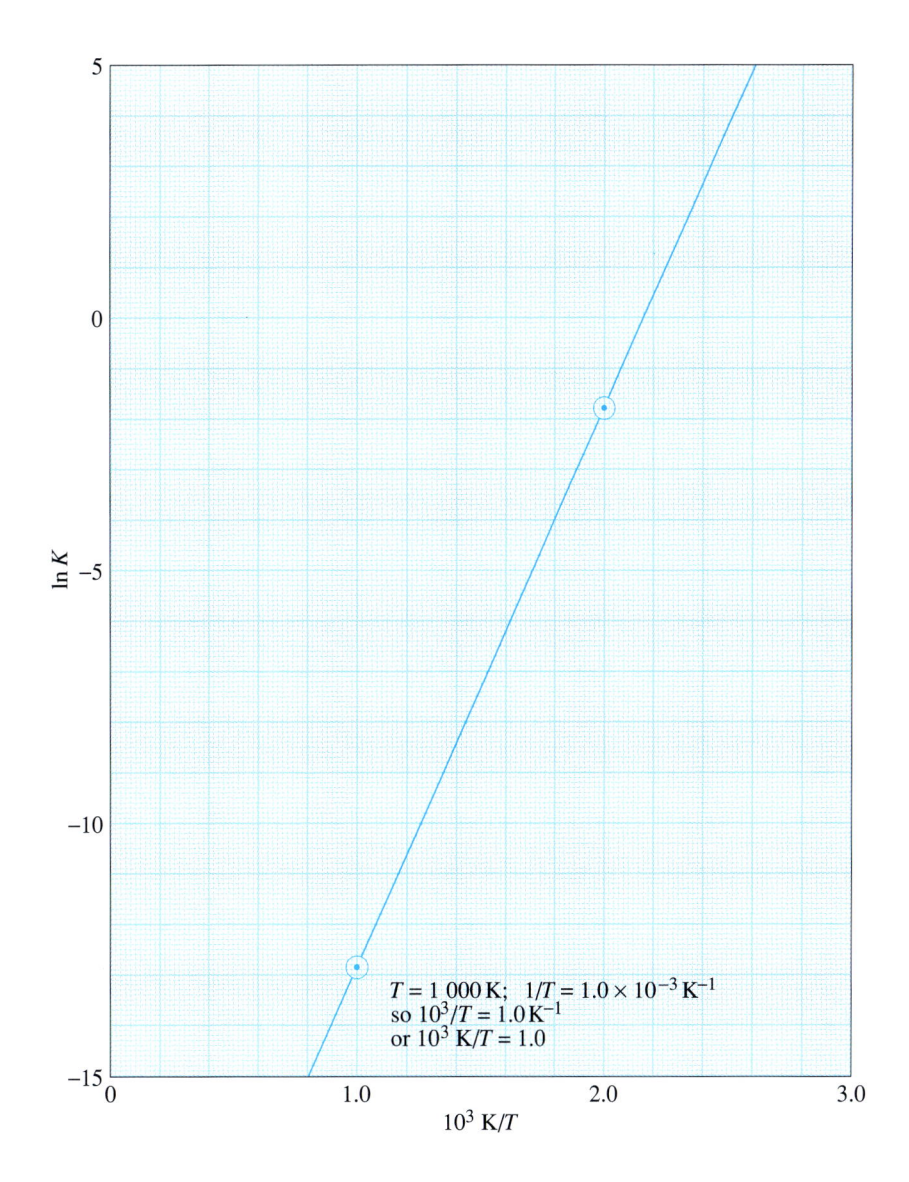

Figure 4 Completed version of Figure 1.

[Graph: ln K vs 10³ K/T, with annotation: T = 1 000 K; 1/T = 1.0 × 10⁻³ K⁻¹ so 10³/T = 1.0 K⁻¹ or 10³ K/T = 1.0]

The justification for joining your two points with a straight line comes from equation 13: if $\Delta H_m^\ominus$ and $\Delta S_m^\ominus$ are assumed constant, then this equation has the same form as that for a straight line ($y = mx + c$), where $y = \ln K$ and $x = (1/T)$. Thus, a plot of $\ln K$ against $1/T$ (y against x) should indeed be a straight line, with slope $m = (-\Delta H_m^\ominus /R)$. In this case, the slope is positive because the *value* of $\Delta H_m^\ominus$ is negative (the reaction is exothermic). For an endothermic reaction, the plot would go the other way, with a negative slope.

SAQ 4 (Objective 2)

(a) Because $\Delta H_m^\ominus < 0$, the reaction is exothermic, so the equilibrium constant should *decrease* with increasing temperature.

(b) In the Second level Inorganic Course, we used an argument very similar to that used here to show that the temperature dependence of $\Delta G_m^\ominus$ for a reaction is determined by the *sign of the entropy change*, $\Delta S_m^\ominus$. From the relation $\Delta G_m^\ominus = \Delta H_m^\ominus - T\Delta S_m^\ominus$ (and assuming the values of $\Delta H_m^\ominus$ and $\Delta S_m^\ominus$ to be independent of temperature), if $\Delta S_m^\ominus$ is positive (as for reaction 6), then $\Delta G_m^\ominus$ should become *more negative* with increasing temperature.

SAQ 5 (revision of the Second Level Inorganic Course and the Science Foundation Course)

The important point is that the form of the equilibrium constant follows directly from the stoichiometry of the balanced reaction *as written* (compare parts (b) and (c), for example):

(a) For reaction 14, $K_{14} = \dfrac{[H^+(aq)][F^-(aq)]}{[HF(aq)]}$

(b) For reaction 1, $K_1 = \dfrac{[NH_3(g)]^2}{[N_2(g)][H_2(g)]^3}$

(c) For reaction 3, $K_3 = \dfrac{[NH_3(g)]}{[N_2(g)]^{1/2}[H_2(g)]^{3/2}}$

Notice that $K_3 = K_1^{1/2}$, so that $\ln K_3 = \tfrac{1}{2} \ln K_1$ and (according to equation 9) $\Delta G_m^\ominus(3) = \tfrac{1}{2} \Delta G_m^\ominus(1)$. This result underlines the dependence of *all* thermodynamic quantities on the stoichiometry of the reaction equation as written.

SAQ 6 (Objectives 2, 3a and 3c)

For reaction 25,

$$K_p = \dfrac{p(\mathbf{2})p(HCl)}{p(\mathbf{1})}$$

From the stoichiometry of equation 25, vinyl chloride (**2**) and HCl are produced in equal amounts. So, providing we start with pure 1,2-dichloroethane (**1**), $p(\mathbf{2}) = p(HCl)$ throughout the reaction, giving the same equilibrium yield y of each product; that is

$$y = \dfrac{p(\mathbf{2})}{p_{tot}} = \dfrac{p(HCl)}{p_{tot}}$$

Thus $p(\mathbf{2}) = p(HCl) = yp_{tot}$

But according to Dalton's law,

$$p(\mathbf{1}) + p(\mathbf{2}) + p(HCl) = p_{tot}$$

so $p(\mathbf{1}) + 2yp_{tot} = p_{tot}$

or $p(\mathbf{1}) = p_{tot}(1 - 2y)$

Substituting these relations into the expression for K_p gives:

$$K_p = \dfrac{(yp_{tot})(yp_{tot})}{(1-2y)p_{tot}} = \left\{\dfrac{y^2}{(1-2y)}\right\} \times p_{tot}$$

In this case, if we start with pure reactant, then 'complete' reaction would yield a mixture containing *equal* amounts of the two products. It follows that the maximum equilibrium yield of vinyl chloride is 50%, that is $y = 0.5$, according to our definition. Notice that this conclusion checks with the expression for K_p: if $y = 0.5$, then $(1 - 2y) = 0$, and the value of K_p would tend to infinity, a situation that corresponds to *complete* conversion into products. (If $y > 0.5$ then K_p would have a negative value!)

(a) An argument similar to that used in the main text then suggests that the equilibrium yield y should decrease with increasing pressure in this case. Once again, this conclusion accords with the predictions of Le Chatelier's principle: as before, it depends on the way p_{tot} appears in the final expression above (that is, on the 'top line') which in turn derives from the stoichiometry of the balanced reaction equation.

(b) The reaction is endothermic: the equilibrium constant, and hence the equilibrium yield of product, should increase with increasing temperature.

SAQ 7 (Objective 3b)

In each case K_p is given (from equation 29) by $K_p = K^{\ominus} \times (p^{\ominus})^{\Sigma}$, and $p^{\ominus} = 1$ bar:

(a) $\Sigma = 1 + 1 - 1 - \frac{1}{2} = \frac{1}{2}$, so $K_p = 4.41 \times 10^{29}$ bar$^{1/2}$ = 4.41×10^{29} (10^5 Pa)$^{1/2}$ = 1.39×10^{32} Pa$^{1/2}$

(b) $\Sigma = 4 + 1 - 2 = 3$, so $K_p = 2.60 \times 10^4$ bar^3 = 2.60×10^4 (10^5 Pa)^3 = 2.60×10^{19} Pa^3

(c) $\Sigma = 1 + 2 - 2 - 3 = -2$, so $K_p = 2.37 \times 10^{-5}$ bar^{-2} = 2.37×10^{-5} (10^5 Pa)$^{-2}$ = 2.37×10^{-15} Pa^{-2}

The important general point to note is that the numerical value of K_p is equal to the corresponding value of $K^{\ominus}$, if, *and only if*, pressures are expressed in bars.

SAQ 8 (Objectives 3b and 4b)

(a) The first step is to substitute the values $y = 0.2$ and $p_{tot} = 1$ bar into equation 24, to give

$$K_p = \left\{ \frac{256}{27} \times \frac{(0.2)^2}{(0.8)^4} \right\} \times \frac{1}{(1 \text{ bar})^2} = 0.925\ 9 \text{ bar}^{-2}$$

So $K^{\ominus} = 0.925\ 9$, $\ln K^{\ominus} = -0.077$.

From the completed version of Figure 1 (Figure 4 in the answer to SAQ 3), the corresponding value of 10^3 K/$T \approx 2.15$, so $1/T \approx 2.15 \times 10^{-3}$ K^{-1} and $T \approx 465$ K.

(b) Now $K_p = 0.9259 \times \dfrac{1}{(10 \text{ bar})^2} = 0.9259 \times 10^{-2}$ bar^{-2}

So $K^{\ominus} = 0.925\ 9 \times 10^{-2}$, $\ln K^{\ominus} = -4.682$.

Again from Figure 4, $1/T \approx 1.73 \times 10^{-3}$ K^{-1}, so $T \approx 578$ K.

As you should expect, raising the overall pressure allows the reaction temperature to be increased for the same equilibrium yield of ammonia.

ANSWER TO EXERCISE

Exercise 1 (Objectives 1–5)

(a) There is an important general point here: *always read the wording of the question carefully*. In this case, it refers to yield, *not* equilibrium yield. Thus, your answer should include reference to *both* thermodynamic factors (which control the equilibrium yield) *and* kinetic factors (which influence the rate of reaction, and hence the actual yield in a given time).

Thermodynamic factors

For reaction 31,

$$CO(g) + 2H_2(g) = CH_3OH(g) \tag{31}$$

$$\Delta H_m^{\ominus}(298.15 \text{ K}) = \Delta H_f^{\ominus}(CH_3OH, g) - \Delta H_f^{\ominus}(CO, g) - 2\Delta H_f^{\ominus}(H_2, g)$$

$$= \{-200.7 - (-110.5) - 0\} \text{ kJ mol}^{-1}$$

$$= -90.2 \text{ kJ mol}^{-1}$$

$\Sigma = +1 - 1 - 2 = -2$

So, the equilibrium yield can be increased by keeping the temperature low and increasing the overall pressure. You may also have suggested removing the product as it is formed (and recirculating the reactants) as a way of displacing the equilibrium to the right.

Kinetic factors

Two ways of increasing the rate mentioned in Section 6 were raising the temperature (but note the conflict with thermodynamic requirements), and using a suitable catalyst (mentioned in part (d)).

(b) The first step is to determine the value of K_p and hence $K^\ominus$, when $y = 0.3$ and $p_{tot} = 1$ bar.

For reaction 31, $K_p = \dfrac{p(CH_3OH)}{p(CO)\{p(H_2)\}^2}$

Since methanol is the desired product, define $y = p(CH_3OH)/p_{tot}$, so $p(CH_3OH) = yp_{tot}$. From Dalton's law, $p(CO) + p(H_2) + p(CH_3OH) = p_{tot}$, so

$p(CO) + p(H_2) = p_{tot} - yp_{tot} = p_{tot}(1 - y)$

But $p(H_2) = 2p(CO)$ throughout (given), so

$3p(CO) = p_{tot}(1 - y)$

$p(CO) = p_{tot}(1 - y)/3$

and $p(H_2) = 2p_{tot}(1 - y)/3$

Substituting in the expression for K_p gives

$$K_p = \dfrac{yp_{tot}}{\left\{\dfrac{p_{tot}(1-y)}{3}\right\}\left\{\dfrac{2p_{tot}(1-y)}{3}\right\}^2}$$

$$= \dfrac{yp_{tot}}{\dfrac{4}{27}(1-y)^3 p_{tot}^3}$$

$$= \left\{\dfrac{27y}{4(1-y)^3}\right\} \times \dfrac{1}{p_{tot}^2}$$

$$= 5.904 \text{ bar}^{-2} \text{ (when } y = 0.3 \text{ and } p_{tot} = 1 \text{ bar)}$$

Then $K^\ominus = 5.904$, $\ln K^\ominus = 1.775\,6$, and the corresponding value of T is calculated from equation 30,

$$T = \dfrac{\Delta H_m^\ominus}{(\Delta S_m^\ominus - R \ln K^\ominus)}$$

In this case, $\Delta H_m^\ominus = -90.2 \text{ kJ mol}^{-1}$ (from part a),

$\Delta S_m^\ominus = S^\ominus(CH_3OH, g) - S^\ominus(CO, g) - 2S^\ominus(H_2, g)$

$= \{239.8 - (197.7) - 2(130.7)\} \text{ J K}^{-1} \text{ mol}^{-1}$

$= -219.3 \text{ J K}^{-1} \text{ mol}^{-1}$

Thus

$$T = \dfrac{-90.2 \times 10^3 \text{ J mol}^{-1}}{\{-219.3 - (8.314 \times 1.775\,6)\} \text{ J K}^{-1} \text{ mol}^{-1}}$$

$= 385 \text{ K}$

Assumptions include: (i) $\Delta H_m^\ominus$ and $\Delta S_m^\ominus$ are assumed constant (the relatively low value of T obtained probably vindicates this assumption); (ii) CO, H_2, CH_3OH *and the mixture* are assumed to behave as ideal gases. A further implicit assumption is taken up in part (d).

(c) At 500 K, and assuming that $\Delta H_m^\ominus$ and $\Delta S_m^\ominus$ are independent of temperature, $\ln K^\ominus$ can be calculated from equation 13, as:

$$\ln K^\ominus (500\,\text{K}) = -\frac{(-90.2 \times 10^3\,\text{J mol}^{-1})}{(8.314 \times 500\,\text{J mol}^{-1})} + \frac{(-219.3\,\text{J K}^{-1}\,\text{mol}^{-1})}{(8.314\,\text{J K}^{-1}\,\text{mol}^{-1})}$$

$$= -4.678\,9$$

So $K^\ominus = 9.29 \times 10^{-3}$, and $K_p = 9.29 \times 10^{-3}\,\text{bar}^{-2}$.

But, from the expressions in part (b), for $y = 0.3$

$$K_p = 5.904 \times \frac{1}{p_{\text{tot}}^2}$$

So

$$9.29 \times 10^{-3}\,\text{bar}^{-2} = 5.904 \times \frac{1}{p_{\text{tot}}^2}$$

and

$$p_{\text{tot}} = \left(\frac{5.904}{9.29 \times 10^{-3}\,\text{bar}^{-2}}\right)^{1/2}$$

$$= 25\,\text{bar}$$

(d) The important general point here is that the analysis in parts (a)–(c) implicitly assumes that formation of methanol is the *only* reaction possible between CO and H_2. But these are the two components of synthesis gas, and it is clear from Figure 3 that a whole range of organic chemicals *could* be formed from this starting material. A thermodynamic analysis like that in Section 6 could be conducted on any one of the reactions included in Figure 3 – or indeed, on other possible reactions between CO and H_2 – but the conclusions would *not* reflect the equilibrium composition of the gas formed by considering all such possible products *simultaneously*. As you might expect, a full thermodynamic analysis of a complex system like this gets rather tricky, but it can be pursued successfully with the aid of a computer.

In this particular case, an interesting consequence of such an analysis is that *most* of the possible alternative products turn out to be *thermodynamically more favourable* than methanol, under virtually all conditions! That methanol is actually produced in this way only serves to highlight a further vital role of industrial catalysts – the *selective* promotion of the formation of the desired product, at the expense of substances that would certainly dominate if the system ever reached complete thermodynamic equilibrium. We shall have more to say about this later on in the Course.

Physical Chemistry
Principles of Chemical Change

BLOCK 2
AN INTRODUCTION TO CHEMICAL KINETICS

CONTENTS

1	INTRODUCTION	5
2	AN EXPERIMENT IN CHEMICAL KINETICS: AN EXAMPLE OF THE EMPIRICAL APPROACH	6
	2.1 Summary of Section 2	10
3	THE RATE OF A CHEMICAL REACTION	11
	3.1 An alternative approach: the extent of reaction	13
	3.2 Summary of Section 3	16
4	EXPERIMENTAL RATE EQUATIONS AND THE ORDER OF A CHEMICAL REACTION	17
	4.1 Reaction order	18
	4.2 Summary of Section 4	19
5	DETERMINING THE ORDER AND RATE CONSTANT FOR A CHEMICAL REACTION	20
	5.1 The differential method	20
	5.2 The integration method	25
	5.3 Summary of Section 5	34
6	THE EFFECT OF TEMPERATURE ON THE RATE OF A CHEMICAL REACTION	35
	6.1 Summary of Section 6	40
7	EXPERIMENTAL METHODS	41
	7.1 Conventional techniques	41
	7.2 Flow methods	44
	7.3 Summary of Section 7	44
8	REACTIONS BETWEEN MOLECULES: ELEMENTARY REACTIONS	45
	8.1 Summary of Section 8	47
9	COLLISION THEORY OF CHEMICAL REACTIONS	48
	9.1 The gas phase	48
	9.2 The solution phase: collisions and encounters	56
	9.3 Summary of Section 9	58
10	TRANSITION STATE THEORY	59
	10.1 Potential energy surfaces	60
	10.2 Transition state theory in the gas phase	67
	10.3 Thermodynamic aspects	72
	10.4 Summary of Section 10	75
OBJECTIVES FOR BLOCK 2		76
SAQ ANSWERS AND COMMENTS		79
ACKNOWLEDGEMENTS		93

1 INTRODUCTION

In Block 1 we reminded you of your background in chemical thermodynamics, and then went on to show how thermodynamic data can be used to predict the equilibrium yield of a product in a chemical reaction. But the analysis, although powerful, had one serious drawback: it said nothing of the time-scale of reaction. This is the province of *chemical kinetics*. This Block and the three that follow are devoted to its study.

Broadly speaking, chemical kinetics is concerned with the measurement and interpretation of the rates of chemical reactions. The measurement of reaction rate and how it depends on factors such as concentration, temperature and the presence of a catalyst, are naturally of considerable interest in both laboratory and industrial practice. For example, the time-scale of an industrial reaction must be economic, though the reaction must not be explosively fast. Beyond practical aspects, important as they are, kinetic investigations are of fundamental interest to the chemist, because they provide one of the most powerful means available for studying what happens *at the molecular level* during a reaction; we call this the **reaction mechanism**.

A reaction mechanism represents a molecular description of how reactants are converted into products during a chemical reaction. To illustrate this, we can consider the reaction between iodide ion, I^-, and hypochlorite ion, ClO^-, which occurs in aqueous solution and is represented by the following chemical equation:

$$I^-(aq) + ClO^-(aq) = IO^-(aq) + Cl^-(aq) \tag{1}$$

Despite the apparent simplicity, all the available information indicates that the reaction does not occur, as its chemical equation might suggest, when iodide ions and hypochlorite ions encounter one another in solution. Rather, a sequence of **simple**, or **elementary**, **reactions**, involving the formation of intermediate species, takes place; a likely sequence is as follows:

$$ClO^- + H_2O \rightleftharpoons HClO + OH^- \tag{2}$$

$$HClO + I^- \longrightarrow HIO + Cl^- \tag{3}$$

$$OH^- + HIO \longrightarrow H_2O + IO^- \tag{4}$$

Solvent water and the intermediate species HClO and HIO are clearly involved in the mechanism of the reaction. You may notice that arrows signs, $\rightleftharpoons$ and $\longrightarrow$, are used in equations 2 to 4: the reason for this notation will be explained in more detail in Section 8. By convention, we do not include the states of the species occurring in a reaction mechanism.

Reaction mechanisms are examined in some detail in Block 3. For the purposes of this Block, it is essential to realize that many reactions – in solution or in the gas phase – do occur by a series of steps, even when the overall stoichiometry of the reaction is apparently very simple. Indeed, a prime objective of many kinetic investigations is the determination of mechanism. An understanding of mechanism not only provides chemical insight, but also may suggest ways of changing the conditions to make a reaction more efficient.

Another aspect of the term 'mechanism' has been provoked by modern experiment and theory: what actually happens *during* each elementary step in a reaction sequence? Attention is now focused on determining how the chemical species involved (be they atoms, molecules, radicals, ions or electrons) approach one another, what their special requirements for reaction are (in terms of energy and orientation, for instance) and how the intimate act of breaking and reforming chemical bonds occurs. One of the main approaches for obtaining this type of information is through the theoretical analysis of models for chemical reactions. Towards the end of this Block we look at two such models: collision theory and transition state theory.

Experiment is at the heart of most kinetic investigation and so this is where we begin our description of chemical kinetics. Our aim is to define the rate of a chemical reaction, and to discover how this rate varies with reaction conditions such as concentration and temperature. This is the fundamental information on which mechanistic studies, either of a chemical or of a theoretical nature, are based.

> **STUDY COMMENT** If you are at all uncertain about how to express and calculate the rate of change of one quantity with respect to another, then this would be an appropriate point to consult Section 3 of the AV Booklet and to listen to the accompanying tape sequence (band 3 on audiocassette 1, *Rate of change and the derivative*).
>
> You should plan to watch video band 1 (*Yields and rates of gaseous reactions*) after you have finished working through Section 3. You may also wish to note that there is a short video sequence associated with the last section in this Block (video band 2, *How do molecules react?*).

2 AN EXPERIMENT IN CHEMICAL KINETICS:
AN EXAMPLE OF THE EMPIRICAL APPROACH

In this Section we examine the results obtained from a typical experiment in chemical kinetics. Our aim is not to analyse these results in detail, but rather to highlight several general points, which will form the basis of material to be discussed in subsequent Sections.

In any kinetic investigation the first, and indispensable, step is to use some form of chemical analysis to establish the **stoichiometry of reaction**. For the example we consider here, the reaction between thiosulfate ion, $S_2O_3^{2-}$, and 1-bromopropane, C_3H_7Br, in a solvent of roughly equal proportions of ethanol and water, the stoichiometry is

$$S_2O_3^{2-} + C_3H_7Br = C_3H_7S_2O_3^- + Br^- \tag{5}$$

(For simplicity, because a mixed solvent is used, the states of the various species that appear in the chemical equation have been omitted.)

The stoichiometry is such that one mole of each reactant is converted into one mole of each product, and there is no evidence for any significant proportions of by-products being formed. In addition, and most importantly, chemical analysis also reveals that there are no *detectable* concentrations of intermediates present during the reaction*. The stoichiometric equation thus applies throughout the *whole* course of the reaction, and the reaction is said to have **time-independent stoichiometry**. It is worth while re-emphasizing that the chemical analysis of a reaction is a very important preliminary activity in any kinetic study, and is one that should never be omitted.

The next step is to follow the reaction as a function of *time*. For a reaction like that in equation 5, which to a good approximation occurs at constant volume, this can be achieved by measuring the concentrations of either reactants or products at different times, generally under *isothermal*, that is constant-temperature, conditions. Of course, a suitable experimental technique must be available. Rather than become involved in experimental detail at this stage, we shall simply quote a set of experimental results (but we shall return to this reaction when we discuss experimental methods in Section 7).

---

* This should *not* be taken to imply that intermediates are not involved in the mechanism of the reaction. The statement should be taken at face value: *within the accuracy of the chemical analysis*, intermediates cannot be detected.

The results of an experiment at 310.7 K are given in Figure 1. It should be clear from the Figure that the initial concentration of thiosulfate ion is greater than that of 1-bromopropane; that is, the thiosulfate ion is in *excess*. The type of plot shown in Figure 1 is called a **kinetic reaction profile**.

The profile shows that the concentrations of reactants and products remain effectively constant after about 5×10^4 seconds (nearly 14 hours) of reaction: thus, the reaction is very close to equilibrium at this point. The concentration of 1-bromopropane decays to a very small value, whereas that of the thiosulfate ion, because it was in excess in the initial reaction mixture, decays to a measurable constant value: the equilibrium position must therefore be well over to the right of equation 5. *From the kinetic viewpoint*, the reaction has gone to **completion**; that is, had the reactants been initially present with equal concentrations, they would both have been virtually completely converted into products.

■ How would you demonstrate, using the kinetic reaction profile in Figure 1, that reaction 5 has time-independent stoichiometry?

▨ If a reaction has time-independent stoichiometry, then the relationship between the concentrations of reactants and the concentrations of products at *any time* during the reaction will be determined by the form of the stoichiometric equation. In the case of reaction 5, it follows that the sum of the concentrations of products and reactants at any time during the reaction should be constant and equal to the sum of the initial concentrations of reactants. This can be confirmed on the Figure.

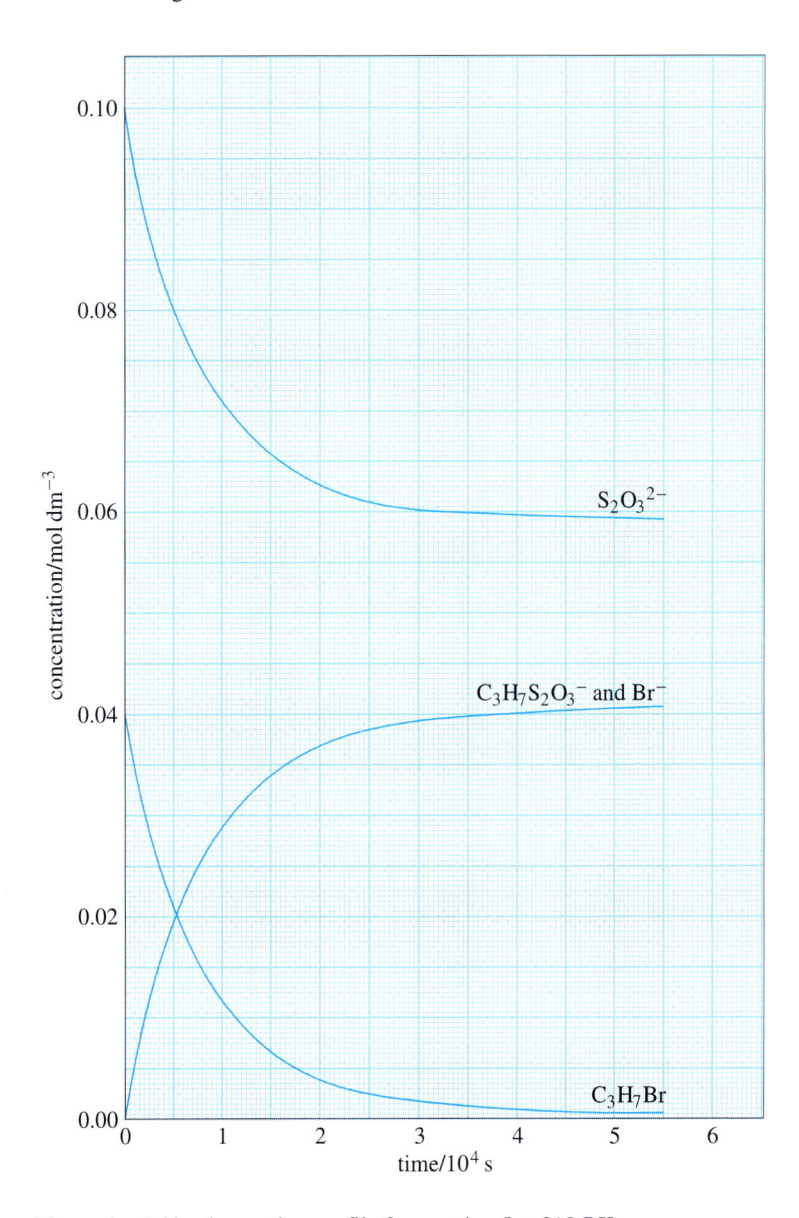

Figure 1 A kinetic reaction profile for reaction 5 at 310.7 K.

The kinetic reaction profile portrays how concentrations change with time during the reaction between thiosulfate ion and 1-bromopropane; however, if this were the only way of communicating kinetic data, life would be tedious indeed. The aim of an **empirical approach** is to describe these changes in the simplest possible mathematical way. Rather than search for an expression that directly relates concentration and time, it turns out simpler, and indeed more informative, to consider how the *rate of reaction* varies as the reaction progresses. So how do we define the rate of reaction?

One definition – but we shall have more to say about this in Section 3 – interprets the rate of reaction as being equal to *the rate of change of concentration with time of a product species*. Thus, if we denote the rate at any time as J, and choose to concentrate on the bromide ion, we can write:

$$J = \frac{d[Br^-]}{dt} \tag{6}$$

■ What are the units of J according to equation 6?

▨ The dimension of J is concentration divided by time and the SI units are $mol\ m^{-3}\ s^{-1}$, although common usage invariably gives them as $mol\ dm^{-3}\ s^{-1}$. Note we could also have written these units as $mol\ l^{-1}\ s^{-1}$ since the litre, l, is defined as $1\ dm^3$. (In this Course we shall use dm^3 in preference to l.)

According to our definition, the instantaneous rate of reaction at any time during the reaction can be determined (as explained in Section 3 of the AV Booklet) from the slope of a tangent to the curve representing the change of bromide ion concentration with time: this is illustrated in Figure 2. The special case corresponding to the start of the reaction is referred to as the **initial rate of reaction**.

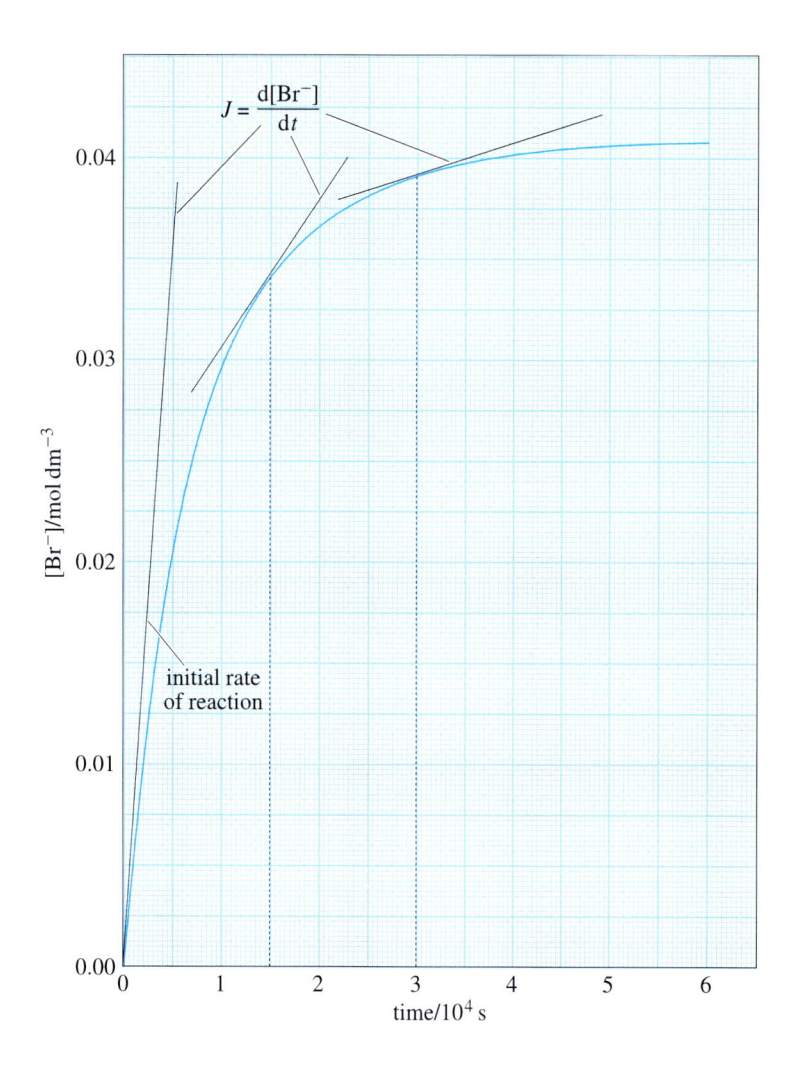

Figure 2 Determination of the initial rate of reaction 5 and the rate of reaction after 1.5×10^4 s and 3.0×10^4 s.

■ What happens to the rate of reaction as time progresses?

■ Figure 2 shows that it decreases. (The slope of the tangent becomes smaller with increasing time.)

The reason for the decrease is the reduction in concentration of *reactants* as the reaction progresses. Is there then a relationship between the rate of reaction and the concentration of reactants? In fact, this is one of the key questions in any kinetic investigation, and much experimental work is devoted to its study. In the case of reaction 5, it turns out that the rate of reaction is proportional to both the concentration of thiosulfate ion and the concentration of 1-bromopropane. This is expressed by the following equation:

$$J = k_R[S_2O_3^{2-}][C_3H_7Br] \tag{7}$$

We refer to this equation as an **experimental rate equation**: it is empirical in the sense that it is the simplest form of mathematical expression that can be found that adequately represents the experimental data. The quantity k_R is called a **rate constant**: it has a value that is *independent* of the concentrations of reactants. *It cannot be overemphasized that there is no link between the stoichiometry of reaction 5 and the form of its experimental rate equation: the latter, as its name implies, is just experimental.*

■ What are the units of k_R according to equation 7?

■ The units of k_R can be calculated from those given by the ratio $J/[S_2O_3^{2-}][C_3H_7Br]$; that is, typically they are $(mol\ dm^{-3}\ s^{-1})/(mol\ dm^{-3})^2$, which simplifies to $dm^3\ mol^{-1}\ s^{-1}$.

Can you suggest a simple way to test whether the experimental rate equation is correct?

One way is to calculate values of $J/[S_2O_3^{2-}][C_3H_7Br]$ at various times during the reaction (from the results in Figures 1 and 2, say) and so test whether, as predicted, this quantity remains constant. The values collected in Table 1 confirm, within experimental uncertainty, that this is the case.

Table 1 Values of the quantity $J/[S_2O_3^{2-}][C_3H_7Br]$ determined at 310.7 K.[a]

time s	$[S_2O_3^{2-}]$ $mol\ dm^{-3}$	$[C_3H_7Br]$ $mol\ dm^{-3}$	J $mol\ dm^{-3}\ s^{-1}$	$J/[S_2O_3^{2-}][C_3H_7Br]$ $dm^3\ mol^{-1}\ s^{-1}$
0	0.100	0.041	6.59×10^{-6}	1.61×10^{-3}
5 000	0.081	0.021	2.67×10^{-6}	1.57×10^{-3}
10 000	0.071	0.011	1.23×10^{-6}	1.57×10^{-3}
15 000	0.066	0.006	0.64×10^{-6}	1.62×10^{-3}
20 000	0.063	0.004	0.41×10^{-6}	1.63×10^{-3}
25 000	0.061	0.002	0.21×10^{-6}	1.72×10^{-3}

[a] As the reaction proceeds, the values of J and $[C_3H_7Br]$ become smaller and more difficult to estimate; consequently, there is an increased uncertainty in the value of $J/[S_2O_3^{2-}][C_3H_7Br]$. The errors are also compounded by the difficulty of drawing 'good' tangents to the curve in Figure 2 to determine J.

The rate of a chemical reaction also depends on *temperature*. In this case, the form of equation 7 suggests that the temperature-dependent factor is the rate constant. This is confirmed in Figure 3, which shows a plot of the experimentally determined rate constant versus temperature. Clearly there is a very marked variation with temperature. (Below 280 K the rate constant does not become zero; it simply has a value some orders of magnitude smaller than those at higher temperatures.)

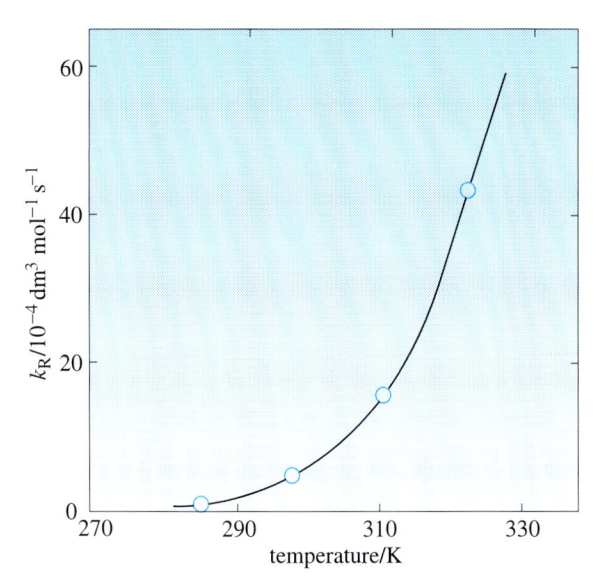

Figure 3 The variation of rate constant with temperature for the reaction between thiosulfate ion and 1-bromopropane in aqueous alcoholic solution.

2.1 Summary of Section 2

In this Section we have looked at the results of a kinetic investigation of the reaction between thiosulfate ion and 1-bromopropane in a solvent of roughly equal proportions of ethanol and water. The main conclusions are:

1 The reaction has a time-independent stoichiometry, which is adequately represented by the following chemical equation:

$$S_2O_3^{2-} + C_3H_7Br = C_3H_7S_2O_3^- + Br^- \tag{5}$$

2 At 310.7 K the reaction effectively goes to completion.

3 The experimental rate equation is of the form:

$$J = \frac{d[Br^-]}{dt} = k_R[S_2O_3^{2-}][C_3H_7Br] \tag{7}$$

4 The rate of reaction, and hence the rate constant, is markedly temperature dependent.

All four observations are important, but the last two raise interesting general questions; for instance

- How is the form of a rate equation established from experimental results?

- What is the best method for determining k_R once a rate equation has been established?

- What forms do the rate equations for other reactions take? Can they be classified in some way?

- Do the rates of all chemical reactions depend markedly on temperature? If so, what is the nature of this temperature dependence?

- Can theoretical models be developed to explain the observed dependence of reaction rate on concentration and temperature?

It is these and other related questions that we shall examine in the remainder of this Block. *In the main we shall concentrate on reactions that have a time-independent stoichiometry and that go to completion,* since in these cases the empirical approach is most easily developed.

There is one obstacle in our path, however. So far, we have defined the rate of reaction somewhat arbitrarily in terms of the rate of change of concentration of a product species. A more general definition is now required. The following SAQ hints at some of the problems in making such a definition, and a more detailed discussion is taken up in the next Section.

SAQ 1 Using only the information given in the kinetic reaction profile in Figure 1, calculate a value for d[Br⁻]/d*t* after 7 500 seconds of reaction. What is the value of d[$C_3H_7S_2O_3^-$]/d*t* after the same time of reaction?

If we had chosen to define the rate of reaction in terms of the rate of change of concentration of a *reactant* species – that is, either thiosulfate ion or 1-bromopropane – what would be the values of d[$S_2O_3^{2-}$]/d*t* and d[C_3H_7Br]/d*t* after 7 500 seconds of reaction?

3 THE RATE OF A CHEMICAL REACTION

The rate of a chemical reaction is a key parameter in any kinetic analysis. It is thus important to define it in a clear and consistent manner. It is worth noting that ambiguities in the scientific literature do occasionally arise because different authors use different definitions, even for the same reaction.

As an example, let us consider how to define the rate of reaction for the gas-phase decomposition of dinitrogen pentoxide, N_2O_5, under *constant-volume* conditions. The decomposition is described by the following equation*:

$$2N_2O_5(g) = 4NO_2(g) + O_2(g) \tag{8}$$

One way to express the rate of reaction would be to write it as the rate of change of concentration of oxygen, that is

$$\frac{d[O_2]}{dt}$$

But what if the reaction had been followed in terms of the change of concentration of dinitrogen pentoxide, or the change in concentration of nitrogen dioxide? Could the rate of reaction be expressed equally well as follows?

$$\frac{d[N_2O_5]}{dt}, \text{ or } \frac{d[NO_2]}{dt}$$

Two problems now arise. First, according to the stoichiometry of the reaction, at any instant the three expressions, d[O_2]/d*t*, d[N_2O_5]/d*t* and d[NO_2]/d*t*, will all have *different* values. This is because the rate of increase of concentration of NO_2 is four times that for oxygen, and twice the rate of decrease of concentration of N_2O_5. Second, the quantity d[N_2O_5]/d*t* is negative, since dinitrogen pentoxide is *consumed* in the reaction (this point also arose in SAQ 1). *However, it would be very helpful to be able to quote a single positive value for the rate of this reaction, and indeed any other reaction, with time-independent stoichiometry.*

As already indicated, equation 8 tells us that the rate of increase of the concentration of NO_2 is four times that for oxygen and twice the rate of decrease of the concentration of N_2O_5. In more mathematical terms

$$\frac{d[O_2]}{dt} = \frac{1}{4}\frac{d[NO_2]}{dt} = -\frac{1}{2}\frac{d[N_2O_5]}{dt} \tag{9}$$

---

* Actually the reaction is slightly more complex than indicated. Depending on reaction conditions, the nitrogen dioxide, NO_2, dimerizes to some extent as soon as it is formed, to give N_2O_4. This association reaction, however, reaches equilibrium very rapidly compared with the rate of the decomposition, so that experimental measurements can easily be corrected for this additional reaction. We shall use this reaction as an example at various stages in this Block; in all cases the experimental measurements will have been amended so that the decomposition is described by equation 8.

where the negative sign ensures that the quantity $-\frac{1}{2}d[N_2O_5]/dt$ has a positive value overall. Any one of the three individual expressions in equation 9 provides an unambiguous definition for the rate of the decomposition reaction; that is,

$$J = \frac{d[O_2]}{dt}, \text{ or } J = \frac{1}{4}\frac{d[NO_2]}{dt}, \text{ or } J = -\frac{1}{2}\frac{d[N_2O_5]}{dt}$$

Our example suggests a more general method of defining the rate of a chemical reaction. To do this, we can write a chemical reaction of *known* stoichiometry in an 'alphabetical' form:

$$aA + bB + ... = pP + qQ + ... \tag{10}$$

where A, B and so on, represent reactants and P, Q and so on, represent products. The numbers $a, b, ...$ and $p, q, ...$ ensure that the equation is balanced. Writing a chemical reaction in this way allows a quantity called the **stoichiometric number** to be introduced. It is given the symbol v_i (Greek letter 'nu'), where the subscript i represents a given species – reactant or product – in the reaction mixture. The stoichiometric number is then defined so that for species A, v_A is $-a$, for species B, v_B is $-b$, for species P, v_P is p, and for species Q, v_Q is q. Hence, the stoichiometric number for a reactant is always **negative**, and that for a product is always **positive**. Normally, chemical reactions are written with the stoichiometric numbers having their *smallest*, but *integral*, values.

■ What are the stoichiometric numbers of N_2O_5, NO_2 and O_2 as given by equation 8?

▨ According to the definition above:

$v_{N_2O_5} = -2$, $v_{NO_2} = 4$ and $v_{O_2} = 1$

In terms of equation 10, the rate of reaction when the volume is constant throughout the reaction – that is, the **rate of reaction at constant volume**, J – can be defined as follows:

$$J = \frac{1}{v_A}\frac{d[A]}{dt} = \frac{1}{v_B}\frac{d[B]}{dt} = \frac{1}{v_P}\frac{d[P]}{dt} = \frac{1}{v_Q}\frac{d[Q]}{dt} \tag{11}$$

■ Is the term $\frac{1}{v_A}\frac{d[A]}{dt}$ positive in value?

▨ Yes! Since A is a reactant, $d[A]/dt$ is negative. But v_A is *defined* to be negative, so that on division a positive value results.

Two important points arise from equation 11:

• The definition of the rate of a chemical reaction will depend on the form in which the balanced chemical equation is written. To avoid ambiguity, this must always be stated in a kinetic study.

• The individual terms can be taken to be equal only if the stoichiometric equation remains valid throughout the course of the reaction; in other words, providing the reaction has time-independent stoichiometry.

Finally, it is worth a small digression to note that partial pressure is often used as a convenient means of expressing concentration when dealing with chemical reactions that occur in the gas phase. In Block 1 you were shown that the partial pressure, $p(A)$, of a gas A in a mixture of gases, each of which obeys the ideal gas equation, can be expressed as:

$$p(A) = \frac{n_A RT}{V} \tag{12}$$

where n_A is the amount of A, V is the volume of the mixture and T is the temperature.

■ What does the quantity n_A/V represent?

▨ It is just the concentration of A, that is [A], in the mixture.

Thus, equation 12 can be written as:

$$p(A) = [A]RT \tag{13}$$

Clearly, as long as neither the temperature nor the volume changes during a reaction, then the partial pressure can be taken to be directly proportional to concentration. The assumption of ideal gas behaviour is usually a good approximation for the conditions under which many kinetic investigations are carried out.

SAQ 2 For each of the reactions 14–16, all of which take place under constant-volume conditions, express the rate of reaction, J, in terms of the rate of change of concentration of both reactants and products in that reaction:

$$2H_2(g) + O_2(g) = 2H_2O(g) \tag{14}$$

$$2NO(g) + 2H_2(g) = N_2(g) + 2H_2O(g) \tag{15}$$

$$BrO_3^-(aq) + 5Br^-(aq) + 6H^+(aq) = 3Br_2(aq) + 3H_2O(l) \tag{16}$$

SAQ 3 If the partial pressure of a gaseous reactant is found to be 1.280×10^4 Pa at some stage during a reaction occurring at 350.0 K, what is the concentration of this reactant expressed in mol dm^{-3}?

3.1 An alternative approach: the extent of reaction

The concept of 'extent' of reaction can also be used to provide a general definition of rate of reaction. As yet, this has not gained widespread acceptance, but nevertheless the underlying ideas are useful in chemical kinetics (and in other branches of physical chemistry, as we shall see in Block 7), and it is for this reason that we outline them here.

The **extent of reaction** is a convenient quantity for describing the progress of a chemical reaction towards its equilibrium position: it is given the symbol ξ (Greek letter 'xi'). In formal terms it is defined as follows:

$$\xi = \frac{n_Y - n_{Y,0}}{v_Y} \tag{17}$$

where Y represents either a reactant or product, n_Y is the amount of substance Y when the reaction has progressed to an extent ξ, $n_{Y,0}$ is the amount of the same substance Y present initially, and v_Y is the stoichiometric number of substance Y in the balanced chemical equation.

■ What is the SI unit of extent of reaction?

▨ Since v_Y is a pure number, the SI unit of ξ relates to the amount of substance: the mole.

The reason for the form of equation 17 is best illustrated by considering a specific example; again let us take the decomposition of dinitrogen pentoxide in the gas phase:

$$2N_2O_5(g) = 4NO_2(g) + O_2(g) \tag{8}$$

Figure 4 shows how the *amounts* of N_2O_5, NO_2 and O_2 change with time when the decomposition takes places at 328.1 K in a sealed vessel with 0.003 mol of dinitrogen pentoxide initially present.

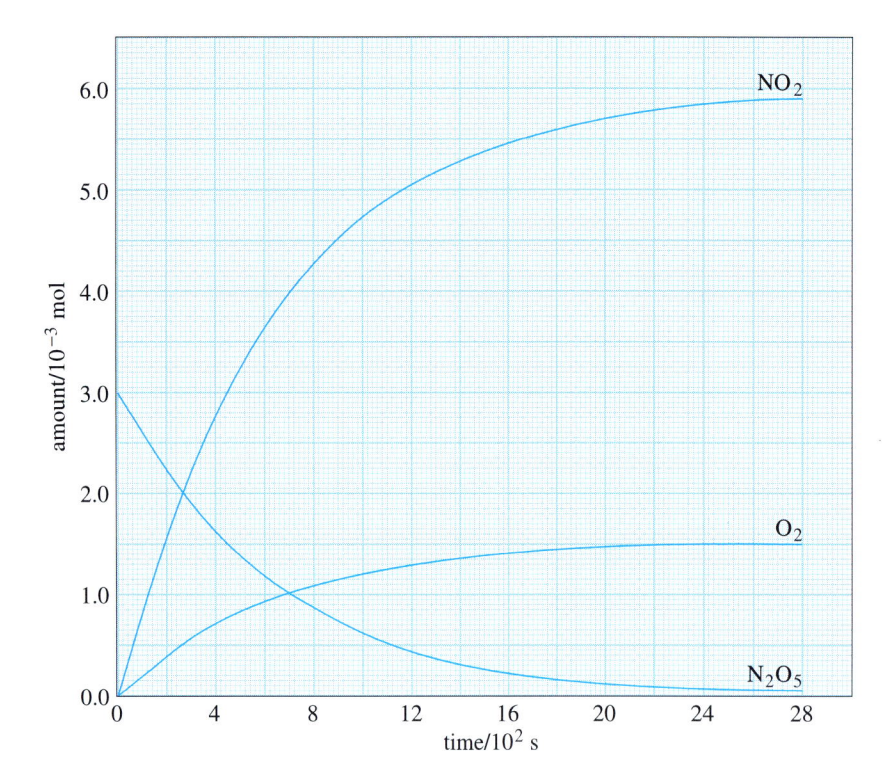

Figure 4 A kinetic reaction profile showing how the amounts of N_2O_5, NO_2 and O_2 change with time for the thermal decomposition of N_2O_5 at 328.1 K. (The initial amount of N_2O_5 was 0.003 mol.)

SAQ 4 Table 2 lists a few of the experimental values used to plot the curves in Figure 4. Calculate the extent of reaction at each time.

Table 2 The thermal decomposition of N_2O_5 at 328.1 K.

$\dfrac{time}{s}$	$\dfrac{n_{N_2O_5}}{10^{-3}\,mol}$	$\dfrac{n_{NO_2}}{10^{-3}\,mol}$	$\dfrac{n_{O_2}}{10^{-3}\,mol}$
0	3.00	0	0
580	1.22	3.56	0.89
1 060	0.59	4.82	1.205
∞	0	6.00	1.50

The answer to SAQ 4 shows that the value of the extent of reaction at a given time is the same, irrespective of whether it is calculated in terms of the amount of reactant consumed or the amounts of products formed. Hence the extent of reaction provides an unambiguous means of describing the 'amount of reaction': it is a positive quantity, which increases as the reaction progresses. It is important to note, however, that the concept of extent of reaction cannot be used unless both of the following conditions are satisfied: (i) the stoichiometry of the reaction is stated, and (ii) the reaction has time-independent stoichiometry. Figure 5 shows a plot of the extent of reaction versus time for reaction 8 at 328.1 K: the Figure incorporates all of the information in Figure 4 but in a more concise form.

The fact that the extent of reaction provides an unambiguous means of describing the progress of a reaction leads to the proposal that the *rate of reaction* be defined as the rate of change of the extent of reaction, that is $d\xi/dt$. However, the units of this rate are typically mol s^{-1}, and this is contrary to the almost universal practice in kinetics of measuring reaction rate in units of (concentration)(time)$^{-1}$. An alternative, but related, recommendation is therefore to define:

$$\text{rate of reaction} = \frac{1}{V}\frac{d\xi}{dt} \qquad (18)$$

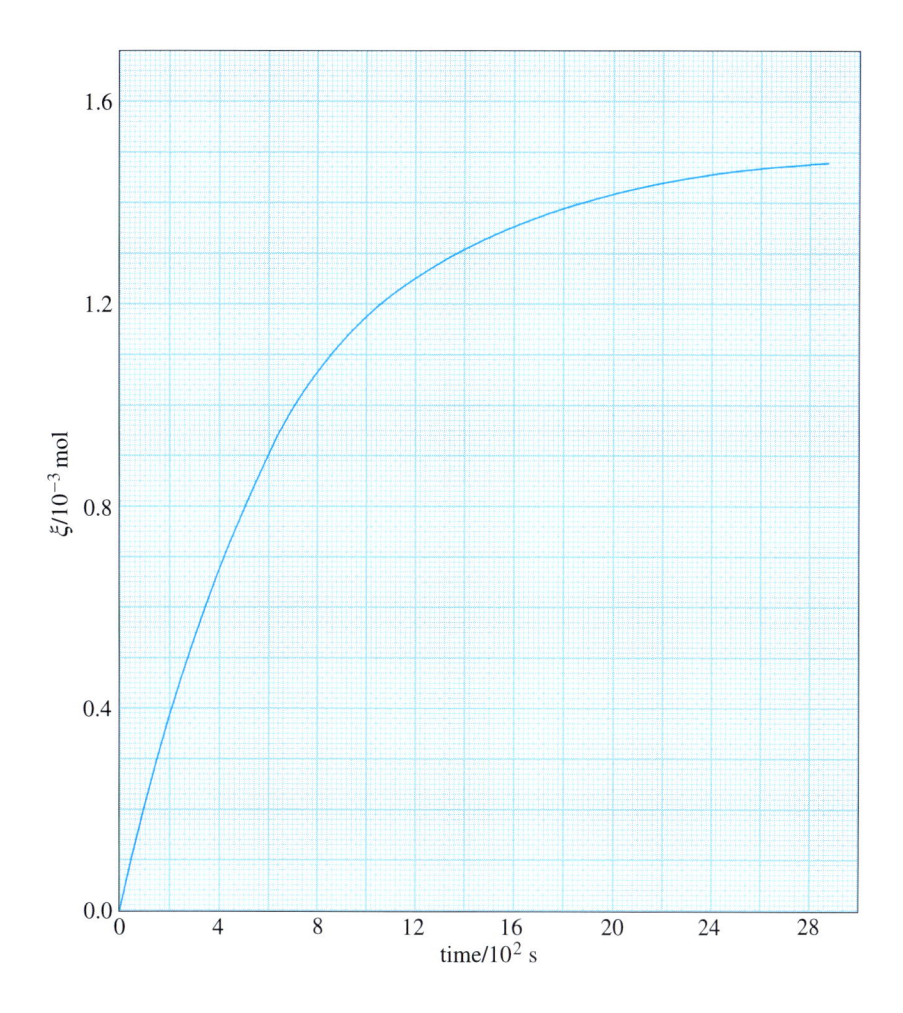

Figure 5 A plot of extent of reaction versus time for the thermal decomposition of N_2O_5 at 328.1 K. (The initial amount of N_2O_5 was 0.003 mol.)

where V is the volume of the reaction mixture. If this volume does not change with time (as, for example, for gas reactions in a sealed vessel, or, to a good approximation, for most solution reactions), then the extent of reaction *per unit volume*, ξ/V, can be found by dividing both sides of equation 17 by V; that is,

$$\frac{\xi}{V} = \frac{1}{\nu_Y}\left(\frac{n_Y}{V} - \frac{n_{Y,0}}{V}\right) \tag{19}$$

The quantity n_Y/V is just the concentration of species Y, that is [Y], when the reaction has progressed to the extent of reaction per unit volume ξ/V, and the quantity $n_{Y,0}/V$ is just the initial concentration of Y, that is $[Y]_0$. The *extent of reaction per unit volume* is also called the **reaction variable** and is given the symbol x, so that:

$$x = \frac{[Y] - [Y]_0}{\nu_Y} \tag{20}$$

It thus follows that a more practical definition of reaction rate for a reaction occurring under constant-volume conditions is dx/dt. According to equation 20 it can then be shown that:

$$\text{rate of reaction at constant volume} = \frac{dx}{dt} = \frac{1}{\nu_Y}\frac{d[Y]}{dt} \tag{21}$$

■ Do you recognize the form of equation 21?

▪ The term on the right-hand side is simply a concise way of writing the general definition of reaction rate given in equation 11. Hence, $J = dx/dt$, as we might expect.

So, to summarize, the rate of change of the reaction variable with time provides a formal means, based on the concept of extent of reaction, for defining the rate of a chemical reaction. In later Sections of this Block we shall make good use of this definition.

SAQ 5 The reaction between 1,2-dibromoethane, $C_2H_4Br_2$, and potassium iodide, KI, in a solution of 99% methanol and 1% water occurs according to the following chemical equation:

$$C_2H_4Br_2 + 3KI = 2KBr + KI_3 + C_2H_4 \tag{22}$$

Express the concentration of potassium iodide at a given time in the reaction in terms of the reaction variable, x, and the initial concentration of potassium iodide, $[KI]_0$.

3.2 Summary of Section 3

1   For a chemical reaction of *known* stoichiometry, taking place under constant-volume conditions, say:

$$aA + bB + \ldots = pP + qQ + \ldots$$

the rate of reaction, J, is defined by the relationship in Box 1.

Box 1

$$J = \frac{1}{\nu_A}\frac{d[A]}{dt} = \frac{1}{\nu_B}\frac{d[B]}{dt} = \frac{1}{\nu_P}\frac{d[P]}{dt} = \frac{1}{\nu_Q}\frac{d[Q]}{dt}$$

where $\nu_A = -a$, $\nu_B = -b$, $\nu_P = p$ and $\nu_Q = q$

2   The individual terms in the expression in Box 1 can be taken to be equal only if the reaction has time-independent stoichiometry.

3   For gas-phase reactions the concentration terms in the expression in Box 1 can in most instances be taken to be proportional to partial pressures.

4   The relationships stated in Box 2 indicate how the extent of reaction ξ and reaction variable x are defined.

Box 2

$$\xi = \frac{n_Y - n_{Y,0}}{\nu_Y}$$

$$x = \frac{\xi}{V} = \frac{[Y] - [Y]_0}{\nu_Y}$$

5   In terms of the reaction variable, the rate of reaction at constant volume J is defined by the relationship in Box 3.

Box 3

$$J = \frac{dx}{dt} = \frac{1}{\nu_Y}\frac{d[Y]}{dt}$$

STUDY COMMENT If you have not done so already, this would be an appropriate point to view video band I (*Yields and rates of gaseous reactions*). You will be able to complete all of the questions associated with the notes for this band when you have worked through Sections 4 and 5.

4 EXPERIMENTAL RATE EQUATIONS AND THE ORDER OF A CHEMICAL REACTION

In Section 2 we described the results of a typical kinetic study of the reaction between thiosulfate ion and 1-bromopropane, and suggested that the reaction could be described by a rate equation of the form:

$$J = k_R[S_2O_3^{2-}][C_3H_7Br] \tag{7}$$

In principle, we would expect every chemical reaction to have a rate equation associated with it, and it is natural to ask about the form of such equations. For instance, are they all of a type in which the rate of reaction is directly related to a product of terms representing the concentrations of reactant species, as in the example above? The experimentally determined rate equations for just a few chemical reactions are given in Table 3.

Glancing through Table 3, you should be quickly convinced of the variety of forms that are found for chemical rate equations. The complex rate equation for the apparently simple reaction between hydrogen and bromine gas could hardly have been anticipated from the stoichiometric equation.

The vital lesson to be drawn from the examples in Table 3 is that:

> The rate equation for a chemical reaction need not necessarily bear any simple relationship to the stoichiometry of the chemical reaction. Of equal importance is the consequence of this statement: chemical rate equations must be determined by experiment.

The form of the experimental rate equation reflects the mechanism of the reaction: we examine this aspect of kinetics in Block 3.

If we had space to extend Table 3 to include more examples, it would become apparent that many of the experimental rate equations do have a common form. This observation leads to the idea of *classifying* chemical reactions in terms of their rate equations.

Table 3 Experimentally determined rate equations[a].

	Overall reaction	Experimentally determined rate equations under isothermal conditions
(a)	$CH_3COOC_2H_5(aq) + OH^-(aq) = C_2H_5OH(aq) + CH_3COO^-(aq)$	$J = k_a[CH_3COOC_2H_5][OH^-]$
(b)	$BrO_3^-(aq) + 5Br^-(aq) + 6H^+(aq) = 3Br_2(aq) + 3H_2O(l)$	$J = k_b[BrO_3^-][Br^-][H^+]^2$
(c)	$H_2O_2(aq) + 2I^-(aq) + 2H^+(aq) = 2H_2O(l) + I_2(aq)$	$J = k_c[H_2O_2][I^-] + k_c'[H_2O_2][I^-][H^+]$
(d)	$H_2(g) + Br_2(g) = 2HBr(g)$	$J = \dfrac{k_d[H_2][Br_2]^{1/2}}{1 + k_d'[HBr]/[H_2]}$

[a] The quantities k_a, k_b, k_c, k_c', k_d and k_d' are constants whose values depend on the reaction conditions. Strictly, we should state these conditions; this is not done here because the examples are for illustrative purposes only.

4.1 Reaction order

If we consider a general chemical reaction:

$$a\text{A} + b\text{B} + c\text{C} = p\text{P} + q\text{Q} + r\text{R} \qquad (23)$$

then it is quite often found that the *experimental* rate equation takes the form:

$$J = -\frac{1}{a}\frac{d[\text{A}]}{dt} = \frac{1}{p}\frac{d[\text{P}]}{dt} = k_R[\text{A}]^\alpha[\text{B}]^\beta[\text{C}]^\gamma \qquad (24)$$

where k_R is the rate constant, and the concentrations of the reactants are [A], [B] and [C]. In this case the **overall order of reaction**, n, is defined by:

$$n = \alpha + \beta + \gamma \qquad (25)$$

The exponent α is referred to as the *order of reaction with respect to reactant A*, β is the order with respect to reactant B, and γ is the order with respect to reactant C; collectively, the quantities α, β and γ are referred to as **partial orders of reaction**.

In some cases, experimental rate equations are found to depend on the concentrations of product species, and some even depend on the concentrations of species that do not explicitly appear in the overall chemical equation (you will meet some examples later in the Course). But, *as long as the experimental rate equation is of a form comparable with that in equation 24*, then the overall order and partial orders can still be defined. In other words, equation 24 can be thought of in broader terms as a rate equation in which [A], [B], [C] and so on represent the concentrations of *any* species – and this includes products – present in the reaction mixture.

■ What are the orders with respect to the individual reactants, and the overall order of reaction for reaction (b) in Table 3?

$$\text{BrO}_3^-(\text{aq}) + 5\text{Br}^-(\text{aq}) + 6\text{H}^+(\text{aq}) = 3\text{Br}_2(\text{aq}) + 3\text{H}_2\text{O}(\text{l}) \qquad (16)$$

▫ The *experimentally* determined rate equation is:

$$J = k_b[\text{BrO}_3^-][\text{Br}^-][\text{H}^+]^2$$

Thus, the order with respect to BrO_3^- is 1, with respect to Br^- it is 1, and with respect to H^+ it is 2. The overall order is therefore, $n = 1 + 1 + 2 = 4$; that is, fourth order overall.

We have already stated that the rate equation for a reaction can be determined only by experiment. It thus follows, *and this cannot be emphasized too strongly*, that the overall order of a reaction is also an experimental quantity. Clearly, for reaction (b) in Table 3 we could not have 'guessed' the form of the rate equation from a knowledge of the chemical equation alone: in general (except in special circumstances which we discuss in Section 8), *there is no direct relationship between the stoichiometric numbers and the partial orders of a reaction*.

For which of the reactions in Table 3 can an overall order of reaction be defined?

The concept of overall order is *meaningless* unless the rate equation is of the general form given by equation 24; that is, a *single* term involving the products of concentrations of species in the reaction mixture, each possibly raised to some power. Hence, reactions (a) and (b) have rate equations for which an order can be defined; reactions (c) and (d) do not.

■ In the presence of ethanol, C_2H_5OH, the following reaction

$$\text{CH}_3\text{COOCH}_3 + \text{CH}_3\text{CH}_2\text{NH}_2 = \text{CH}_3\text{CONHCH}_2\text{CH}_3 + \text{CH}_3\text{OH} \qquad (26)$$

has an experimental rate equation of the form:

$$J = k_R \frac{[\text{CH}_3\text{CH}_2\text{NH}_2]^{3/2}[\text{CH}_3\text{COOCH}_3]}{[\text{C}_2\text{H}_5\text{OH}]^{1/2}} \qquad (27)$$

What are the orders with respect to the individual species that appear in the rate equation, and the overall order of reaction?

■ The order with respect to $CH_3CH_2NH_2$ is $\frac{3}{2}$, with respect to CH_3COOCH_3 it is 1, and with respect to C_2H_5OH it is $-\frac{1}{2}$. The overall order is thus $n = \frac{3}{2} + 1 - \frac{1}{2} = 2$; that is, second order overall.

This example makes three important points:

- Even though ethanol does not appear in the overall chemical equation, the experimental rate equation is of such a form that the order with respect to ethanol can be defined.

- The order with respect to a species in a reaction mixture may be a positive, negative or fractional number: it *must* be determined by experiment.

- The overall order is simply a means of classifying a chemical reaction. The bald statement that the reaction between methyl ethanoate, CH_3COOCH_3, and ethylamine, $CH_3CH_2NH_2$, in the presence of ethanol is second order overall, tells us little about the chemistry involved. It is the *partial orders* of reaction that are more informative.

4.2 Summary of Section 4

1 The rate equation for a chemical reaction need not necessarily bear any simple relationship to the stoichiometry of the overall chemical equation: it must be determined by experiment.

2 The procedure for working out the overall order of a reaction is summarized in Box 4.

Box 4

Write down the *experimental* rate equation. If, and only if, it has the general form:

$$J = k_R[A]^\alpha[B]^\beta[C]^\gamma \ldots \quad (24)$$

where A, B, C, and so on are any species present in the reaction mixture, then the overall order is defined as:

$$n = \alpha + \beta + \gamma + \ldots \quad (25)$$

The quantities α, β and γ, which are partial orders with respect to the individual species in the reaction mixture, may be positive, negative or fractional numbers.

SAQ 6 For each of the experimentally determined rate equations given in Table 4, state the orders of reaction with respect to the individual species that appear in the rate equation, and the overall order of reaction.

Table 4 Experimental rate equations.

Reaction	Experimental rate equation
(a) $2NO(g) + O_2(g) = 2NO_2(g)$	$J = k_a[NO]^2[O_2]$
(b) $CO(g) + Cl_2(g) = COCl_2(g)$	$J = k_b[CO][Cl_2]^{3/2}$
(c) $H_2O_2(aq) + 2Fe^{2+}(aq) + 2H^+(aq) = 2Fe^{3+}(aq) + 2H_2O(l)$	$J = k_c[Fe^{2+}][H_2O_2]$
(d) $2O_3(g) = 3O_2(g)$	$J = k_d \dfrac{[O_3]^2}{[O_2]}$
(e) $(CH_3)_3CCl(aq) + OH^-(aq) = (CH_3)_3COH(aq) + Cl^-(aq)$	$J = \dfrac{k_e[(CH_3)_3CCl]}{1 + k_e'[Cl^-]}$

5 DETERMINING THE ORDER AND RATE CONSTANT FOR A CHEMICAL REACTION

So far we have seen a number of examples of experimental rate equations, but as yet we have said little about how these equations are determined from experimental results. As explained in Section 2, the first step in any kinetic study is to determine the stoichiometry of a reaction. This is then followed by measurement of the changes in concentrations of either reactants or products as a function of time, generally under isothermal conditions. The procedure for establishing a rate equation from such measurements is based on an amalgam of two methods: the **direct** or **differential method** and the **integration method**. In this Section we examine both of these methods in some detail. In the main, we focus attention on reactions that are either first order or second order, since these are relatively easy to handle. In practice, chemical reactions, particularly those used in industry, are often more complicated, but the principles underlying their investigation remain the same.

5.1 The differential method

This method has been used for many years since it was first suggested by the Dutch physical chemist J. H. van't Hoff in 1884, and for this reason it is sometimes named after him. Consider again the example discussed in Section 3 – the thermal decomposition of dinitrogen pentoxide:

$$2N_2O_5(g) = 4NO_2(g) + O_2(g) \tag{8}$$

The first step is to propose a **plausible rate equation.** To do this, the *simplest* proposal is made. Generally*, this means that we assume that the rate of reaction depends *only* on the concentrations of reactant species. So, for reaction 8, a plausible rate equation will be

$$J = k_R[N_2O_5]^\alpha \tag{28}$$

where the partial order, α, is the same as the overall order of reaction, n.

Next we must establish whether this assumed rate equation is consistent with the experimental results and, if so, determine the overall order. In the differential method approach this means determining the rate of reaction, J, at different instants in time as the reaction progresses, and hence at different concentrations of N_2O_5 in the reaction mixture.

■ How is the rate of reaction, J, at a given instant in time determined?

□ The rate of reaction at a given instant in time is determined from a kinetic reaction profile:

$$J = -\frac{1}{2}\frac{d[N_2O_5]}{dt}, \text{ or } J = \frac{1}{4}\frac{d[NO_2]}{dt}, \text{ or } J = \frac{d[O_2]}{dt}$$

However, as your experience in SAQ 1 may confirm, it is difficult to do this with any accuracy: this is, in fact, a disadvantage of the differential method. Given values of J as the reaction progresses, how can we now test whether the assumed rate equation, namely equation 28, fits the experimental results?

---

* It is possible that there may be some form of experimental evidence available that indicates a dependence of the rate of a reaction on a product species or 'another species' in the reaction mixture. In this Course, if it is necessary, you will *always* be given such additional information before being asked to suggest a plausible rate equation.

One method would be to guess a value of the order, say $\alpha = 1$, and then check whether the quantity $J/[N_2O_5]$ was a constant. If not, another value of α would be tried, and so on. Trial-and-error procedures are never very satisfactory, and so we should try to devise a better approach. In particular we should try to rewrite equation 28 in the form of a straight line, and so use a graphical approach.

Can you see how to do this?

Taking logarithms of equation 28 gives the result

$$\begin{aligned}\log J &= \log (k_R[N_2O_5]^\alpha) \\ &= \log k_R + \log ([N_2O_5]^\alpha) \\ &= \log k_R + \alpha \log [N_2O_5]\end{aligned} \quad (29)$$

You should appreciate that since logarithms of both sides of equation 28 are taken, it is immaterial whether these are base ten (log) or natural, base e (ln), logarithms. Notice also that we have written log (quantity) – that is, $\log [N_2O_5]$ – rather than the more cumbersome, but strictly correct, log (quantity/unit) – that is, $\log ([N_2O_5]/\text{mol dm}^{-3})$. In future usage, we shall adopt the former practice, except when plotting graphs or giving tables of information, when the strictly correct procedure will always be followed.

Equation 29 describes a straight line, so that:

- a plot of $\log J$ versus $\log [N_2O_5]$ should be linear if the assumed form of the rate equation is correct;
- the slope of the straight line will be given by α, thereby providing a value for the order of reaction with respect to N_2O_5.

As explained in Section 3, partial pressure is often used as a convenient means of expressing concentration when dealing with chemical reactions that occur in the gas phase. Thus in Figure 6 the logarithm of the rate of reaction, expressed in units of Pa s^{-1}, is plotted against the logarithm of the partial pressure of dinitrogen pentoxide, $p(N_2O_5)$, for the reaction at 328.1 K. The plot is linear and the straight line has a slope of 0.96 with an estimated uncertainty of ±0.05; that is, the value is very close to unity. The order of reaction does not have to be a whole number, but it often is, and so the value of α is almost certainly 1, and the rate equation takes the form:

$$J = -\frac{1}{2}\frac{d[N_2O_5]}{dt} = k_R[N_2O_5] \quad (30)$$

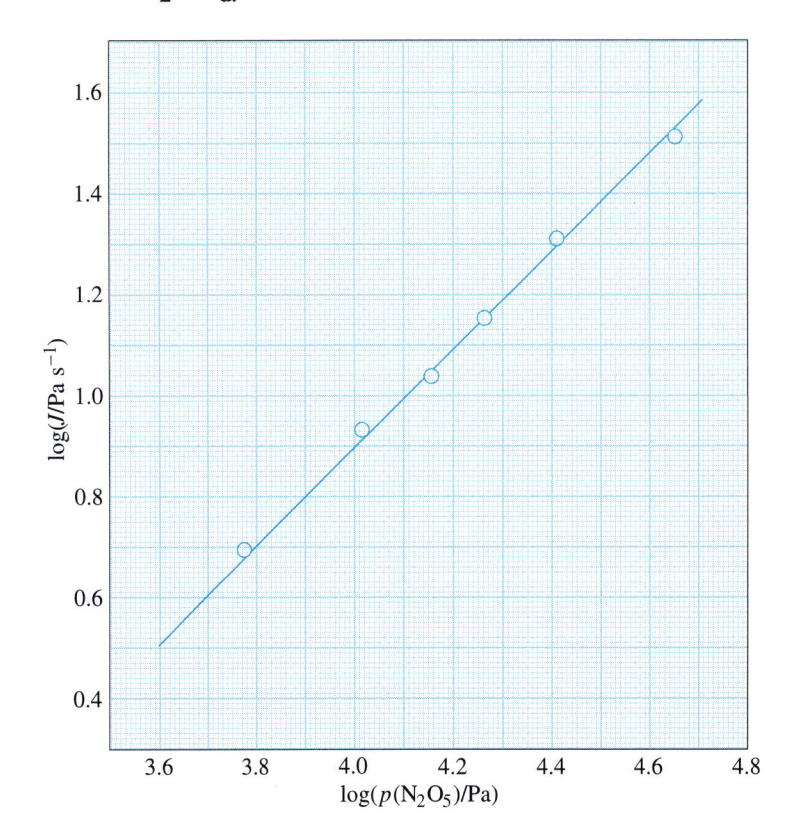

Figure 6 A plot of $\log (J/\text{Pa s}^{-1})$ versus $\log (p(N_2O_5)/\text{Pa})$ for the thermal decomposition of dinitrogen pentoxide at 328.1 K.

■ How would you determine the *rate constant* for the decomposition reaction?

■ Equation 30 suggests one method: if the concentration of N_2O_5 at a particular time is known, k_R can be calculated by determining the corresponding value of J.

To increase the reliability of the determination, the calculation can be repeated at different times during the reaction so that an average value of k_R can be arrived at. The disadvantage of this method – as we have already indicated – is that values of J are difficult to measure with great accuracy. Alternatively, returning to Figure 6 and equation 29, the intercept when $\log(p(N_2O_5)/Pa)$ is zero gives a value for $\log k_R$; but again this method would involve a long, and probably unreliable, extrapolation. In fact, having established the form of a rate equation, it is better to use a different approach to determine the rate constant: you will learn about such an approach in Section 5.2.

The values plotted in Figure 6 were taken from the kinetic reaction profile for a 'single kinetic run'. An alternative strategy is to perform a series of experiments at different *initial* concentrations of reactant, $[N_2O_5]_0$, and then to evaluate, as accurately as possible, the initial rates of reaction, J_0. In these circumstances, equation 29 takes the form:

$$\log J_0 = \log k_R + \alpha \log [N_2O_5]_0 \tag{31}$$

This is the basis of the **initial rate method**, which is discussed more fully in the next Section.

The differential method provides a general and convenient graphical technique for establishing the form and order of a chemical rate equation. To apply the method to a reaction for which the rate is thought to depend on more than one concentration term, requires the use of an **isolation technique**. In essence, the technique involves *isolating* in turn the contribution of each reagent by arranging that all other reagents are in *large excess*, such that their concentrations remain virtually unchanged during the course of reaction. Normally, this means at least a ten-fold, but more preferably a forty-fold, excess in concentration compared with the initial concentration of the reactant to be isolated. On occasions this may not be possible, for reasons such as limited solubilities of reagents.

To illustrate the isolation technique, we can look again at the reaction we considered in Section 2, the reaction between thiosulfate ion and 1-bromopropane:

$$S_2O_3^{2-} + C_3H_7Br = C_3H_7S_2O_3^- + Br^- \tag{5}$$

A plausible rate equation for this reaction would be:

$$J = k_R[S_2O_3^{2-}]^\alpha [C_3H_7Br]^\beta \tag{32}$$

Taking logarithms of this expression gives

$$\log J = \log k_R + \alpha \log [S_2O_3^{2-}] + \beta \log [C_3H_7Br] \tag{33}$$

which is *not* a linear equation, since $\log J$ depends on both $\log [S_2O_3^{2-}]$ and $\log [C_3H_7Br]$. However, if the experimental conditions are arranged such that the thiosulfate ion is in *large excess* (that is, in terms of initial concentrations, $[S_2O_3^{2-}]_0 \gg [C_3H_7Br]_0$), then equation 32 can be simplified. Because, in relative terms, the concentration of thiosulfate ion will hardly change as the reaction progresses, it is reasonable to replace the term $[S_2O_3^{2-}]$ by the term $[S_2O_3^{2-}]_0$; that is,

$$J = k_R[S_2O_3^{2-}]_0^\alpha [C_3H_7Br]^\beta \tag{34}$$

or, more concisely,

$$J = k_R'[C_3H_7Br]^\beta \tag{35}$$

where $k_R' = k_R[S_2O_3^{2-}]_0^\alpha$. The isolation technique has simplified the rate equation: it now has a **pseudo-order** and the quantity k_R' is called a **pseudo-order rate constant**. The order with respect to 1-bromopropane, that is β, can now be determined in a straightforward graphical manner. If the experiment is repeated, but with 1-bromopropane in excess, the order with respect to thiosulfate ion, α, can be found.

SAQ 7 In aqueous solution, methyl ethanoate, CH_3COOCH_3, reacts to produce ethanoic acid, CH_3COOH, and methanol, CH_3OH:

$$CH_3COOCH_3(aq) + H_2O(l) = CH_3COOH(aq) + CH_3OH(aq) \tag{36}$$

For low concentrations of methyl ethanoate (less than $0.1\ mol\ dm^{-3}$), the experimentally determined rate equation is of the form:

$$J = k_R[CH_3COOCH_3] \tag{37}$$

Can you suggest one reason why the experimental rate equation does not contain a term involving the concentration of water?

5.1.1 The method of initial rates: a case study

A number of chemical reactions produce products that either decompose or interfere in some way with the progress of reaction. A method of investigating such reactions is to measure the initial rates of reaction, since 'by definition' these depend only on the concentrations of species present initially in the reaction mixture. Although such an approach provides information on only the initial stages of a reaction, it is none the less important in obtaining a broader understanding of a particular reaction. Experimentally, changes in *product* concentration are used to determine the initial rate: high analytical accuracy would be required to measure relatively small changes in the initially high concentration of a reactant.

To illustrate the method we shall take as an example the conversion of the hypochlorite ion, ClO^-, into the hypoiodite ion, IO^-, in alkaline solution, that is, in the presence of the hydroxide ion, OH^-. The overall reaction at 298 K is:

$$I^-(aq) + ClO^-(aq) = IO^-(aq) + Cl^-(aq) \tag{1}$$

The reaction does not go to completion and is complicated by the unstable nature of the hypoiodite ion, IO^-. However, this *product* species is stable enough for its concentration to be determined during the initial stages of reaction.

■ A series of preliminary experiments revealed that the rate of reaction depended on the concentrations of ClO^-, I^- and OH^-, even though the last species does not appear in the chemical equation. Suggest a plausible rate equation.

▪ A plausible rate equation would be:

$$J = k_R[ClO^-]^\alpha[I^-]^\beta[OH^-]^\gamma \tag{38}$$

The initial rate of reaction will depend only on the initial concentrations of species that appear in the rate equation, that is:

$$J_0 = k_R[ClO^-]_0^\alpha\,[I^-]_0^\beta\,[OH^-]_0^\gamma \tag{39}$$

If J_0 were measured using a variety of initial concentrations, then a series of equations involving the four unknowns, k_R, α, β and γ, would result: these could be solved, but it is tedious.

Can you suggest how to simplify the investigation?

An isolation technique can be used. However, because initial rates are to be measured, all that is necessary is to perform experiments in which the *initial* concentrations of selected reagents are kept constant – it is *not* necessary to have reagents in excess. Three distinct groups of experiments can be performed in which the initial rates of reaction are measured as a function of:

A   the initial concentration of OH^-, while the initial concentrations of ClO^- and I^- are held fixed;

B   the initial concentration of I^-, while the initial concentrations of ClO^- and OH^- are held fixed;

C   the initial concentration of ClO^-, while the initial concentrations of I^- and OH^- are held fixed.

■ Can you see how this simplifies the analysis of the kinetic data? Try to write a rate equation for, say, the first series of experiments.

▨ If $[ClO^-]_0$ and $[I^-]_0$ are held constant, then equation 39 simplifies to:

$$J_0 = k_R'[OH^-]_0^\gamma \qquad (40)$$

where $k_R' = k_R[ClO^-]_0^\alpha [I^-]_0^\beta$ and is a pseudo-order rate constant.

■ How then would you determine the order, γ, with respect to the hydroxide ion?

▨ Taking logarithms of equation 40

$$\log J_0 = \log k_R' + \gamma \log [OH^-]_0 \qquad (41)$$

shows that a plot of $\log J_0$ versus $\log [OH^-]_0$ will be a straight line with slope γ.

Exactly the same reasoning underlies the other two series of experiments; it is confirmed by the logarithmic plots in Figure 7.

■ What then is the form of the experimental rate equation, and what is the overall order of reaction?

▨ The fact that the plots in Figure 7 are all linear confirms the form of rate equation suggested in equation 38. The slopes of the straight lines are -1.02, 0.98 and 0.99, respectively. These values are very close to whole numbers, so the rate equation for the initial stages of reaction is most likely to be:

$$J = k_R \frac{[ClO^-][I^-]}{[OH^-]} \qquad (42)$$

and the overall order of reaction is $n = 1 + 1 - 1 = 1$; that is, the reaction is first order overall.

■ What happens to the rate of reaction as the initial concentration of hydroxide ion is increased?

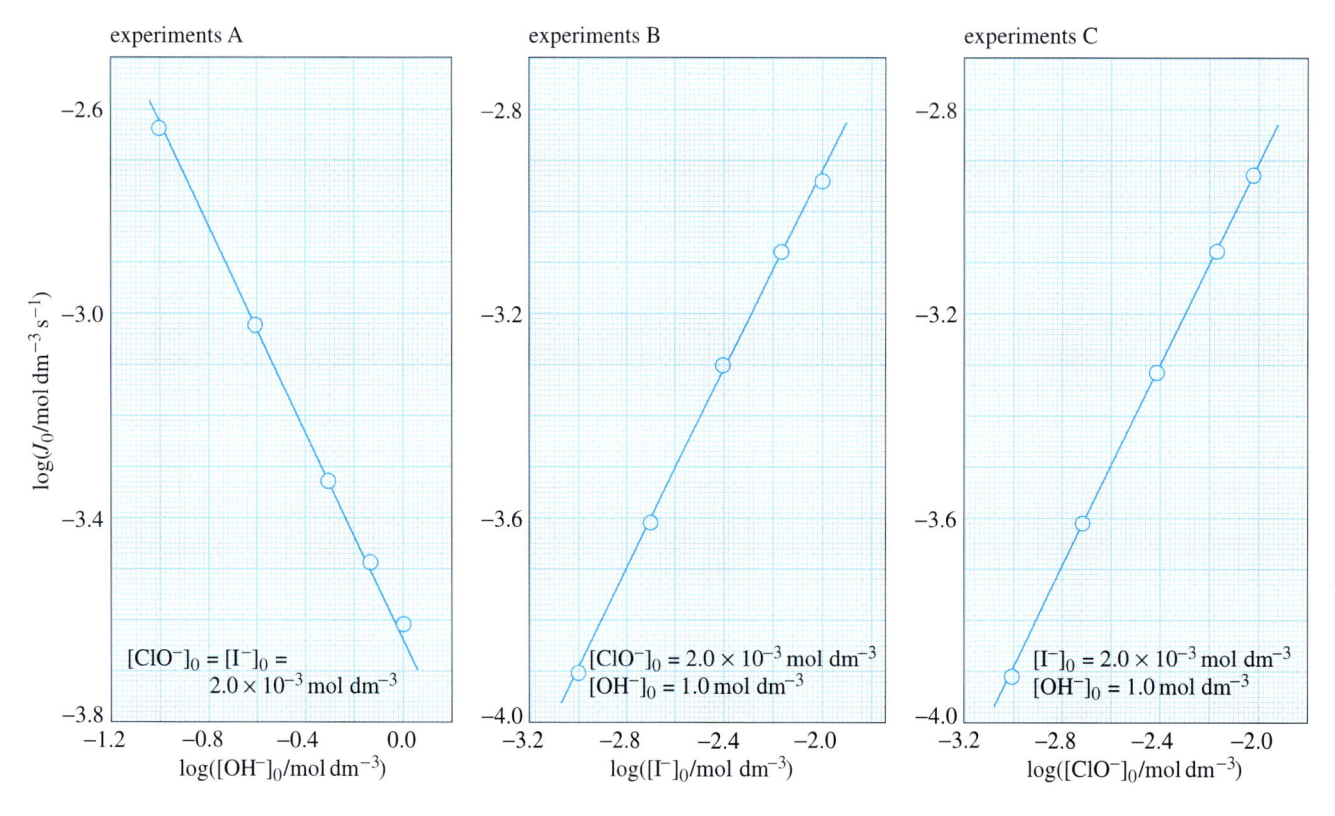

Figure 7 Logarithmic plots for the reaction between iodide ion and hypochlorite ion in alkaline solution at 298 K: (A) with the initial concentrations of ClO^- and I^- held constant; (B) with the initial concentrations of ClO^- and OH^- held constant; (C) with the initial concentrations of I^- and OH^- held constant. (Source of data: Y. T. Chia and R. E. Connick, *J. Phys. Chem.*, 1959)

■ It is slowed down: for this reason the hydroxide ion is called an **inhibitor**.

Finally, it is worth noting that it is not always necessary to use a graphical method in conjunction with initial rate data. In particular, for cases in which the partial orders of reaction are integral numbers, it is often possible to deduce this by simple inspection of initial rate measurements that have been made using an isolation technique. Such a 'kinetic puzzle' is illustrated in the following SAQ.

SAQ 8 For the reaction

$$IO_3^-(aq) + 5I^-(aq) + 6H^+(aq) = 3I_2(aq) + 3H_2O(l) \tag{43}$$

the initial rates of reaction vary with initial concentrations at 298 K as given in Table 5. Suggest a form for the rate equation. What are the orders with respect to the individual species that appear in the rate equation, the overall order of reaction and the rate constant at 298 K?

Table 5 Initial rate data at 298 K for reaction 43.

$\dfrac{[H^+]_0}{10^{-3} \text{ mol dm}^{-3}}$	$\dfrac{[I^-]_0}{10^{-4} \text{ mol dm}^{-3}}$	$\dfrac{[IO_3^-]_0}{10^{-4} \text{ mol dm}^{-3}}$	$\dfrac{J_0}{10^{-9} \text{ mol dm}^{-3} \text{ s}^{-1}}$
2.0	4.0	0.37	7.1
2.0	4.0	0.74	14.2
2.0	4.0	1.48	28.4
1.0	4.0	1.48	7.1
2.0	2.0	1.48	7.1
1.0	1.0	1.48	0.44

STUDY COMMENT Before studying the next section, you may wish to turn to Section 4 of the AV Booklet and listen to the accompanying tape sequence (band 4 on audiocassette 1). The title is *Integration*, and it deals with the background to this mathematical technique and, in particular, it shows how integrated rate equations can be derived. However, it must be emphasized that in this Course you will not be expected to carry out any integrations for yourself; our main concern will be with how integrated rate equations are used in practice. ■

5.2 The integration method

The experimental kinetic information for a reaction consists of a series of measurements of the concentrations of either reactants or products at various times during a reaction. As you have seen, these measurements can be used to determine instantaneous values of the reaction rate, so that the differential method can be used to establish the form of the rate equation and hence the reaction order. *Once the order of reaction is known*, the final step is to determine the value of the rate constant at a particular temperature: the integration method is particularly suited to this purpose. But also, as you will see, the method is more powerful than this single application might suggest.

5.2.1 Integrated rate equations

The rate equation for a chemical reaction is mathematically in the form of a *differential equation* and as such may be integrated directly (in simple cases, at least). This forms the basis of the procedure for determining the rate constant. Two steps are involved:

● Write down the rate equation in a form that is suitable for integration;

● Integrate the rate equation and decide on a graphical method for determining the rate constant from experimental measurements.

It is important that you are familiar with some of the 'procedural' details of these two steps because these will help you in using integrated rate equations in practice. *However, you are not expected to be able to carry out the integration process itself.*

To begin with, we consider a reaction involving only a single reactant, which has a first-order experimental rate equation.

First-order reaction involving a single reactant

In Section 5.1 we established that the thermal decomposition of dinitrogen pentoxide,

$$2N_2O_5(g) = 4NO_2(g) + O_2(g) \tag{8}$$

has an experimental rate equation of the form:

$$J = k_R[N_2O_5] \tag{30}$$

If we express J in terms of the rate of change of concentration of N_2O_5, that is,

$$J = -\frac{1}{2}\frac{d[N_2O_5]}{dt} \tag{44}$$

then the resultant differential equation is

$$-\frac{1}{2}\frac{d[N_2O_5]}{dt} = k_R[N_2O_5] \tag{45}$$

This is written in terms of just one variable, the concentration of N_2O_5 at any time in the reaction, $[N_2O_5]$, and may be integrated directly. The result is:

$$\ln[N_2O_5]_0 - \ln[N_2O_5] = 2k_R t \tag{46}$$

where $[N_2O_5]_0$ represents the initial concentration of N_2O_5. Equation 46 is called an **integrated rate equation**.

■ According to equation 46, how can the rate constant, k_R, be determined?

▨ Equation 46 can be rearranged so that

$$\ln[N_2O_5] = -2k_R t + \ln[N_2O_5]_0 \tag{47}$$

The form of this equation is such that a plot of $\ln[N_2O_5]$, or $\ln(p(N_2O_5))$, against time, t, will give a straight line with a slope that is related to the rate constant.

A plot of $\ln(p(N_2O_5))$ versus time for the decomposition of N_2O_5 at 328.1 K is given in Figure 8.

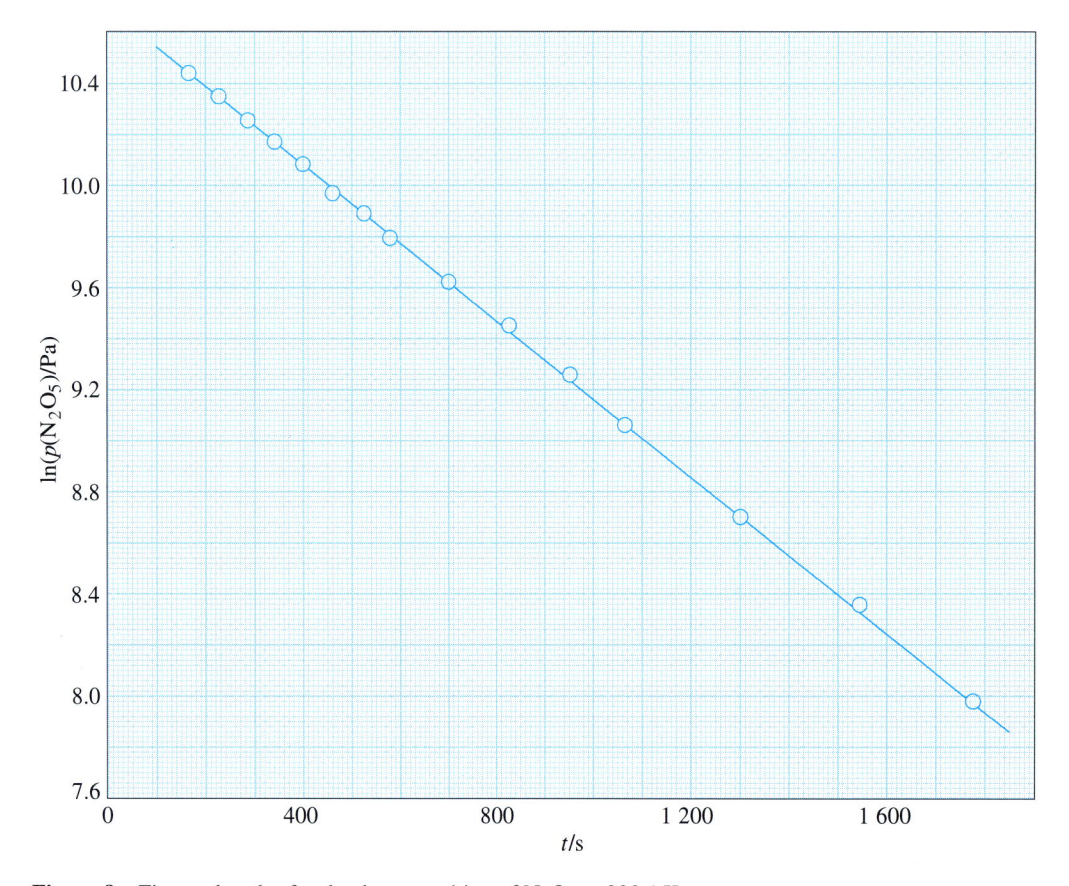

Figure 8 First-order plot for the decomposition of N_2O_5 at 328.1 K.

BLOCK 2 AN INTRODUCTION TO CHEMICAL KINETICS

STUDY COMMENT Before moving on, you should try SAQ 9.

SAQ 9 Use the information in Figure 8 to determine a value of the rate constant, k_R, for the decomposition of N_2O_5 at 328.1 K. What are the units of this rate constant?

It should be noted that equation 46 can be rearranged as follows:

$$\ln\left\{\frac{[N_2O_5]_0}{[N_2O_5]}\right\} = 2k_R t \qquad (48)$$

so that on taking the antilogarithm (or inverse natural logarithm)*,

$$\frac{[N_2O_5]_0}{[N_2O_5]} = \exp(2k_R t) \qquad (49)$$

The latter equation can be written in a more convenient form as follows:

$$[N_2O_5] = [N_2O_5]_0 / \exp(2k_R t)$$
$$= [N_2O_5]_0 \exp(-2k_R t) \qquad (50)$$

Thus, the concentration of N_2O_5 decays exponentially with time: *this behaviour is quite general for a reactant in a first-order reaction.*

SAQ 10 Calculate the length of time it will take for 90% of the initial concentration of N_2O_5 present in a sealed reaction vessel to decompose at 328.1 K. You should use the value of the rate constant you determined in SAQ 9 in your calculation. [*Hint* When 90% of the N_2O_5 has decomposed, the remaining concentration can be expressed as $[N_2O_5] = 0.1[N_2O_5]_0$.]

If the initial concentration of N_2O_5 was 0.01 mol dm^{-3}, what would be the concentration of the product, NO_2, after this time of reaction? [*Hint* The reaction variable can be used in this part of the calculation.]

Second-order reaction involving two reactants

This is a more complex case, and it must be said straightaway that the isolation technique is more often than not used to simplify the study of such reactions. However, consider as an example the reaction between 1,2-dibromoethane, $C_2H_4Br_2$, and potassium iodide, KI, in a solution of 99% methanol and 1% water, according to the chemical equation:

$$C_2H_4Br_2 + 3KI = KI_3 + 2KBr + C_2H_4 \qquad (22)$$

The experimental rate equation is:

$$J = k_R[C_2H_4Br_2][KI] \qquad (51)$$

so that it is second order overall. In differential form we can write:

$$-\frac{d[C_2H_4Br_2]}{dt} = k_R[C_2H_4Br_2][KI] \qquad (52)$$

Do you see a problem with integrating this equation?

The right-hand side of the equation involves two variables, $[C_2H_4Br_2]$ and $[KI]$, although of course these are related by the stoichiometry of the chemical equation. *However, to integrate the equation, it must be recast into a form involving only one variable.* A formal way of achieving this is to use the reaction variable, x, which we introduced in Section 3.1 and also hinted that you should use in SAQ 10. Quantities of interest are expressed in terms of this variable in Table 6.

---

* The exponential of a quantity x can be written as either e^x or $\exp(x)$; the latter notation is mostly used when the exponential of a quantity involving several terms is to be indicated (as is done in equation 49). You should remember also that $1/\exp(x)$ is equal to $\exp(-x)$.

Table 6 Use of the reaction variable for reaction 22.

Substance, Y	ν_Y	$[Y]_0$	$[Y]$
$C_2H_4Br_2$	-1	$[C_2H_4Br_2]_0$	$[C_2H_4Br_2]_0 - x$
KI	-3	$[KI]_0$	$[KI]_0 - 3x$
KI_3	$+1$	0	x
KBr	$+2$	0	$2x$
C_2H_4	$+1$	0	x

Recalling that the rate of reaction can be expressed as dx/dt, we can thus rewrite equation 52 as:

$$\frac{dx}{dt} = k_R ([C_2H_4Br_2]_0 - x)([KI]_0 - 3x) \tag{53}$$

In this equation the initial concentrations of reactants – that is, $[C_2H_4Br_2]_0$ and $[KI]_0$ – are constants and so there is just one variable, x. As might be anticipated, the integrated form of this differential equation is quite complicated:

$$\ln\left\{\frac{([C_2H_4Br_2]_0 - x)}{([KI]_0 - 3x)}\right\} = (3[C_2H_4Br_2]_0 - [KI]_0)k_R t + \ln\left(\frac{[C_2H_4Br_2]_0}{[KI]_0}\right) \tag{54}$$

None the less, if you look at the equation carefully, you will see that it describes a straight line: a plot of the left-hand side versus time is linear, as Figure 9 shows, with slope equal to $(3[C_2H_4Br_2]_0 - [KI]_0)k_R$. Note that the left-hand side of the equation is just the logarithm of the ratio of the reactant concentrations at any time in the reaction.

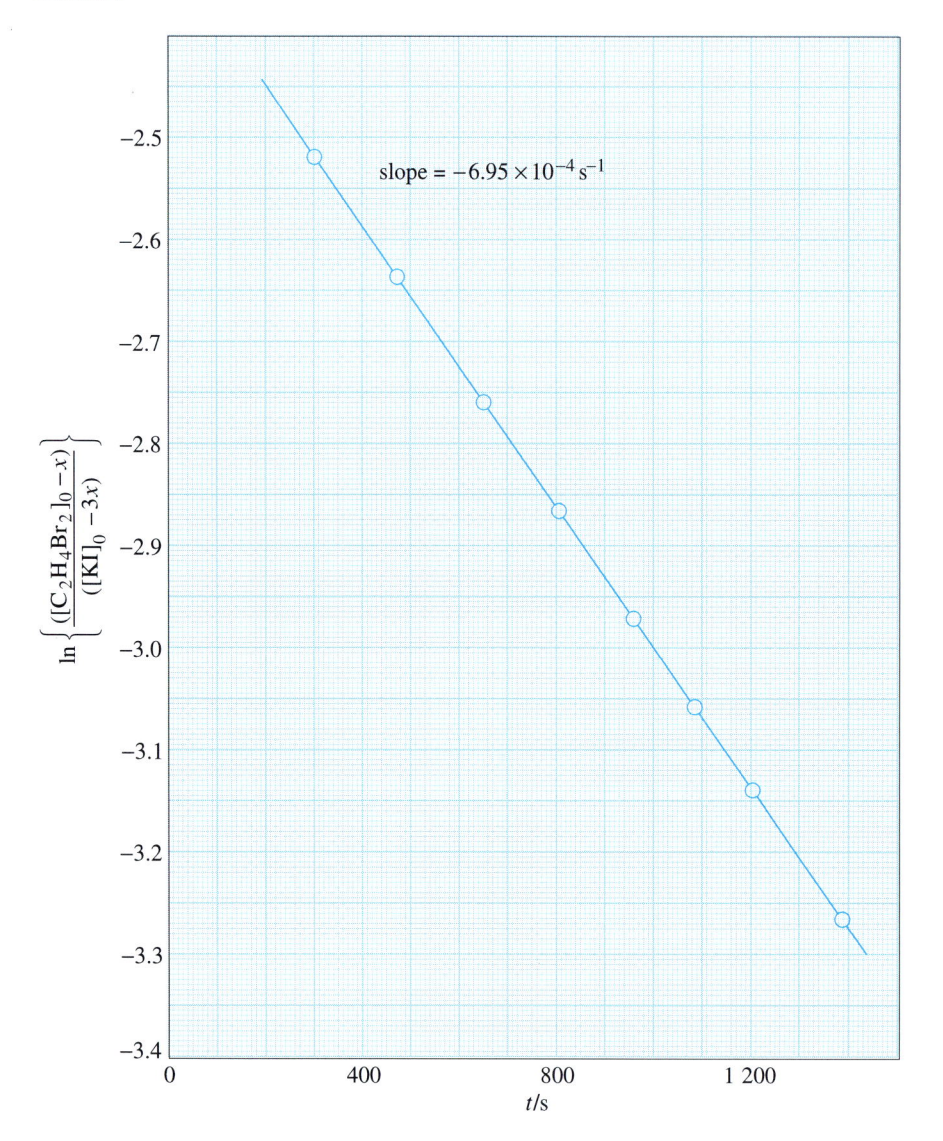

Figure 9 Second-order plot for the reaction between $C_2H_4Br_2$ and KI at 332.9 K. As shown on the plot, the slope of the straight line is $-6.95 \times 10^{-4}\,\text{s}^{-1}$.

- For the experiment in Figure 9, the initial concentrations of $C_2H_4Br_2$ and KI were 0.02 mol dm^{-3} and 0.2 mol dm^{-3}, respectively. What is the rate constant, k_R, for the reaction at 332.9 K?

- The slope of the straight line in Figure 9 is related to the second-order rate constant by the equation:

$$\text{slope} = (3[C_2H_4Br_2]_0 - [KI]_0)k_R \qquad (55)$$

It follows therefore that:

$$k_R = \frac{(\text{slope})}{(3[C_2H_4Br_2]_0 - [KI]_0)} = \frac{-6.95 \times 10^{-4}\,\text{s}^{-1}}{-0.14\,\text{mol dm}^{-3}}$$
$$= 4.96 \times 10^{-3}\,\text{dm}^3\,\text{mol}^{-1}\,\text{s}^{-1}$$

This example highlights two important features of the use of the reaction variable:

- It provides a formal way of writing the rate equation in terms of a single variable, together with the initial concentrations of reactants.

- It can be used in the calculation of reactant concentrations that appear in the integrated rate equation from, say, experimentally determined values of *product* concentrations. For instance, the reaction between 1,2-dibromoethane and potassium iodide is followed in practice by monitoring the concentration of KI_3 as the reaction progresses. As Table 6 shows, the concentration of KI_3 provides a direct value for the reaction variable.

5.2.2 The tabulation of integrated rate equations

In the previous Section we dealt specifically with the integration of two types of rate equation. In practice, many different forms of rate equation may occur, but fortunately the integrated form of only a limited number of these is ever required. Table 7 lists the standard forms of first- and second-order rate equations, their integrated forms, and the simplest functions for plotting graphically in order to determine the rate constant: all the entries refer to reactions of the general type

$$aA + bB = \text{products} \qquad (56)$$

Table 7 Differential rate equations and integrated rate equations for general reactions of the type: $aA + bB = $ products.

Typical reactions[a]	Reaction order	Rate equation	Integrated rate equation	
$aA = $ products	1	$-\dfrac{1}{a}\dfrac{d[A]}{dt} = k_R[A]$	$\ln [A]_0 - \ln [A] = ak_R t$ graphical plotting: $\ln [A]$ versus t	(57)
$aA = $ products	2	$-\dfrac{1}{a}\dfrac{d[A]}{dt} = k_R[A]^2$	$\dfrac{1}{[A]} - \dfrac{1}{[A]_0} = ak_R t$ graphical plotting: $\dfrac{1}{[A]}$ versus t	(58)
$A + B = $ products	2	$-\dfrac{d[A]}{dt} = k_R[A][B]$	$\ln\left(\dfrac{[A]}{[B]}\right) = ([A]_0 - [B]_0)k_R t + \ln\left(\dfrac{[A]_0}{[B]_0}\right)$ graphical plotting: $\ln\left(\dfrac{[A]}{[B]}\right)$ versus t	(59)
$aA + bB = $ products	2	$-\dfrac{1}{a}\dfrac{d[A]}{dt} = -\dfrac{1}{b}\dfrac{d[B]}{dt}$ $= k_R[A][B]$	$\ln\left(\dfrac{[A]}{[B]}\right) = (b[A]_0 - a[B]_0)k_R t + \ln\left(\dfrac{[A]_0}{[B]_0}\right)$ graphical plotting: $\ln\left(\dfrac{[A]}{[B]}\right)$ versus t	(60)

[a] This column lists reactions that *could possibly* give rise to the rate equations indicated. In general, *any* of the rate equations could arise from the single reaction referred to in the heading to this Table.

The concentrations [A] and [B] in Table 7 are the concentrations at a given time in the reaction, and the concentrations [A]$_0$ and [B]$_0$ are those at the start of the reaction. Notice that three different forms of integrated rate equation are given for reactions that are second order overall. The first of these, equation 58, is for a reaction that involves only a *single* reactant: the second, equation 59, is a specific example (that is, with $a = b = 1$) of the general equation, equation 60.

SAQ 11 Equation 60 in Table 7 gives a general form for a second-order integrated rate equation. Show how the specific form of integrated rate equation given in equation 54 for the reaction between 1,2-dibromoethane and potassium iodide can be derived from this general form.

In some kinetic studies, the reactants are mixed in **stoichiometric proportions**. If we consider the case of a reaction involving just two reactants, as described by equation 56, then the ratio of the initial concentrations of reactants A and B, when they are mixed in stoichiometric proportions, will be given by

$$\frac{[A]_0}{[B]_0} = \frac{a}{b} \qquad (61)$$

that is, they are in the same ratio as the 'balancing coefficients' in the stoichiometric equation.

It turns out that in these particular circumstances there is a problem in using a second-order integrated rate equation to determine a rate constant.

■ Can you see what this is?

▪ In equation 60, the quantity $(b[A]_0 - a[B]_0)$ will be zero.

This 'mathematical hitch' is resolved as follows. Suppose that the experimental rate equation is second order; that is,

$$-\frac{1}{a}\frac{d[A]}{dt} = k_R[A][B] \qquad (62)$$

■ If the reaction mixture initially contains reactants in stoichiometric proportions, what will be the value of the ratio [A]/[B] at any time in the reaction?

▪ If the reaction has time-independent stoichiometry, then the ratio will be the same as for the initial concentrations of reactants; that is, a/b.

Hence, we can write:

$$b[A] = a[B] \qquad (63)$$

so that on substitution into equation 62

$$-\frac{1}{a}\frac{d[A]}{dt} = k_R[A] \times \left(\frac{b}{a}[A]\right)$$

$$= \frac{k_R b}{a}[A]^2 \qquad (64)$$

■ Looking at Table 7, do you recognize the form of this equation?

▪ It is of a similar form to that for a second-order reaction involving only a *single* reactant; that is, equation 58.

The integrated form of equation 64 is found by comparison with equation 58 in Table 7, as follows:

$$\frac{1}{[A]} - \frac{1}{[A]_0} = a\left(\frac{k_R b}{a}\right)t = k_R bt \qquad (65)$$

A plot of $1/[A]$ versus time will yield a straight line with slope equal to $k_R b$: thus the rate constant can be determined.

With the information in Table 7, you are now in a position to calculate directly, by a graphical method and hence more accurately, the rate constants for a variety of first- and second-order reactions.

SAQ 12 Given the information in Section 2, in particular that in Table 1 and Figure 1, what graphical method is most suitable for determining the rate constant for the reaction between thiosulfate ion and 1-bromopropane at 310.7 K?

SAQ 13 Using the data in Table 1, what value do you obtain for the rate constant for the reaction between thiosulfate ion and 1-bromopropane at 310.7 K? (Make sure that you state the units for the rate constant in your answer.)

> **STUDY COMMENT** Make sure that you try SAQ 14 before moving on, because it provides a useful summary of some of the factors to be considered when dealing with second-order reactions.

SAQ 14 The reaction between peroxodisulfate ion, $S_2O_8^{2-}$, and iodide ion, I^-, in aqueous solution:

$$S_2O_8^{2-}(aq) + 2I^-(aq) = 2SO_4^{2-}(aq) + I_2(aq) \tag{66}$$

has an experimentally determined rate equation of the form:

$$\frac{d[I_2]}{dt} = k_R[S_2O_8^{2-}][I^-] \tag{67}$$

The reaction is followed as a function of time by monitoring the concentration of iodine, I_2, formed. Describe, *without giving experimental detail*, how you would determine the rate constant for this reaction. (Assume that the peroxodisulfate ion is initially in slight excess.)

5.2.3 The determination of reaction order

So far we have seen how the integration method provides a convenient and accurate graphical method for determining the rate constant for a chemical reaction, *once the order has been established*. But conversely, the fact that an integrated rate expression provides a good fit to the plot of experimental concentration versus time data is confirmation that the order has been correctly determined. This strongly suggests another use for integrated rate equations: the comparison of a plot of concentration versus time with various integrated rate expressions to find the 'best fit' will determine the most appropriate rate equation and hence the reaction order. In fact, this strategy is commonly used in practice, particularly if suitable computer software is available. There is, however, one pitfall with this approach, as we shall see below.

Suppose we consider a gas-phase reaction involving a single reactant A for which the kinetic reaction profile has been measured:

$$aA = \text{products} \tag{68}$$

■ What are the simplest forms of first-order and second-order rate equations for this reaction?

first order: $\quad -\dfrac{1}{a}\dfrac{d[A]}{dt} = k_R[A] \tag{69}$

second order: $\quad -\dfrac{1}{a}\dfrac{d[A]}{dt} = k_R[A]^2 \tag{70}$

In principle, the two rate equations may be tested by plotting the experimental results in the form suggested by the integrated rate equations in Table 7; that is, for a first-order reaction, $\ln p(A)$ is plotted against time, and for a second-order reaction, $1/p(A)$ is plotted against time. Appropriate plots, using the results from a real experiment, are given in Figure 10.

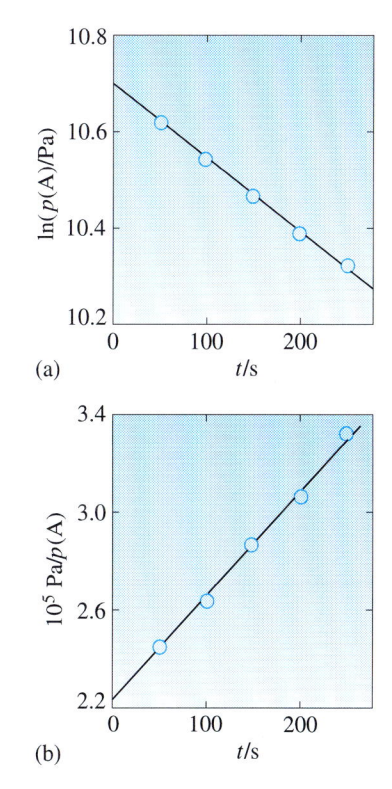

Figure 10 (a) A first-order plot; (b) a second-order plot.

Is the reaction first order or second order?

If we take into account the fact that there will be some scatter due to experimental error, then *both* graphs give comparable straight lines; it would be injudicious to select one of the rate equations. What has gone wrong?

In fact, the data used in plotting the graphs are taken from experiments on the thermal decomposition of N_2O_5 at 328.1 K.

If you compare the data in Figure 10a with those in Figure 8 (Section 5.2.1), can you now see the problem?

The comparison reveals that the difference between the two graphs is in the *time-scale*. In fact, in Figure 8 the data extend to about 94% reaction, whereas in Figure 10a the data are plotted to only about 32% reaction. The situation is made even clearer if the data in Figure 8 are replotted as a second-order plot (see Figure 11): the pronounced curvature emphatically shows that the reaction is *not* second order.

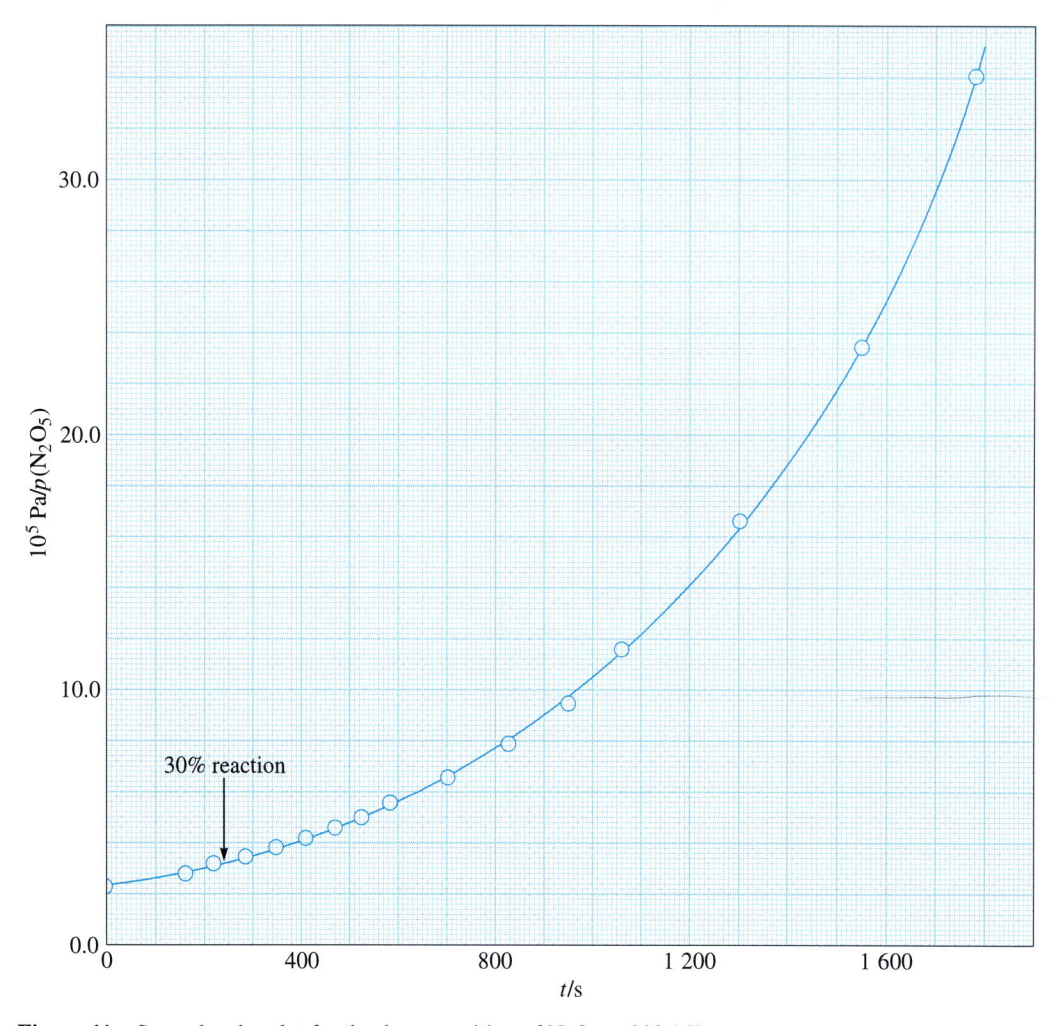

Figure 11 Second-order plot for the decomposition of N_2O_5 at 328.1 K.

The warning is very clear:

> Unless a reaction has been followed to near completion, and certainly to an extent of more than 50%, the integration method can be very misleading in the determination of reaction order.

5.2.4 Reaction half-life

An expression for the **reaction half-life**, $t_{\frac{1}{2}}$, of a chemical reaction can be determined from the appropriate integrated rate equation: the half-life is defined as the time it takes for the concentration of a reactant to fall to half of its initial value. Apart from providing a convenient means of communicating the time-scale of a reaction, we shall see that the measurement of half-life provides a useful diagnostic test for a first-order reaction.

The integrated rate equation for a first-order reaction involving only a single reactant (see Table 7) is:

$$\ln [A]_0 - \ln [A] = ak_R t \tag{57}$$

so that on rearrangement

$$\ln \left\{ \frac{[A]_0}{[A]} \right\} = ak_R t \tag{71}$$

■ The half-life is defined as the time, $t_{\frac{1}{2}}$, taken for the concentration of A to fall from $[A]_0$ to $\frac{1}{2}[A]_0$. Use equation 71 to find an expression for $t_{\frac{1}{2}}$.

□ Clearly, when $[A] = \frac{1}{2}[A]_0$, equation 71 can be written as

$$\ln \left\{ \frac{[A]_0}{\frac{1}{2}[A]_0} \right\} = ak_R t_{\frac{1}{2}} \tag{72}$$

or

$$\ln 2 = ak_R t_{\frac{1}{2}} \tag{73}$$

The reaction half-life is then given by

$$t_{\frac{1}{2}} = \frac{\ln 2}{ak_R} = \frac{0.693}{ak_R} \tag{74}$$

> Notice the implication of equation 74: the reaction half-life for a first-order reaction is *independent* of the initial concentration of reactant.

■ What is the half-life for the thermal decomposition of N_2O_5 at 328.1 K? (You calculated k_R at 328.1 K for this reaction in SAQ 9 as $k_R = 7.635 \times 10^{-4}$ s^{-1}.)

$$2N_2O_5(g) = 4NO_2(g) + O_2(g) \tag{8}$$

□ Using equation 74, with $a = 2$,

$$t_{\frac{1}{2}} = \frac{0.693}{2 \times 7.635 \times 10^{-4} \text{ s}^{-1}} = 454 \text{ s}$$

Notice also that the half-life could have been determined from the kinetic reaction profile in Figure 4 (Section 3.1). (The fact that the profile for N_2O_5 is plotted in terms of amount rather than concentration does not affect the argument.)

Since the half-life of a first-order reaction is independent of the concentration, it follows that, during the first half-life of a reaction, the concentration of a reactant falls to $[A]_0/2$, in two half-lives it falls to $[A]_0/4$, and in three half-lives it falls to $[A]_0/8$. You should check this for yourself on Figure 4. *This characteristic behaviour is often used as a quick preliminary test for the order of a reaction based on the kinetic reaction profile.* Alternatively, arbitrary concentrations may be chosen on the profile and the times taken for these concentrations to fall to half their values determined: if the time intervals are constant, it is a good indication that the reaction is first order. This behaviour for a chemical reaction directly parallels that for the nuclear reaction of radioactive decay, which is a classic example of a first-order decay process.

For reactions of order other than unity, the reaction half-life becomes dependent on initial concentration.

■ Derive an expression for the half-life of a second-order reaction involving only one reactant.

▪ The appropriate integrated rate equation (see Table 7) is:

$$\frac{1}{[A]} - \frac{1}{[A]_0} = ak_R t \tag{58}$$

Hence, at $t = t_{\frac{1}{2}}$, $[A] = [A]_0/2$, and therefore

$$\frac{2}{[A]_0} - \frac{1}{[A]_0} = ak_R t_{\frac{1}{2}} \tag{75}$$

or

$$\frac{1}{[A]_0} = ak_R t_{\frac{1}{2}} \tag{76}$$

The reaction half-life is then given by

$$t_{\frac{1}{2}} = \frac{1}{ak_R [A]_0} \tag{77}$$

According to equation 77, for a second-order reaction involving only one reactant, the reaction half-life increases linearly with the *reciprocal* of the initial concentration of reactant: *the higher the initial concentration, the shorter the half-life*.

The half-lives of other more complicated reactions can be determined from appropriate integrated rate equations: their use is limited, however, and we shall not derive them here.

SAQ 15 The half-life of the gas-phase reaction between iodine pentafluoride, IF_5, and fluorine, F_2, which produces iodine heptafluoride, IF_7, was investigated at 328.8 K:

$$IF_5(g) + F_2(g) = IF_7(g) \tag{78}$$

In a series of experiments, the results of which are in Table 8, the initial concentrations of iodine pentafluoride and fluorine were kept the same, that is $[IF_5]_0 = [F_2]_0$. Determine whether the rate equation is first, second, or some other order.

Table 8 Reaction half-life data for reaction 78 at 328.8 K. In each experiment, the initial concentrations of IF_5 and F_2 were such that $[IF_5]_0 = [F_2]_0$.

initial concentrations mol dm^{-3}	$t_{\frac{1}{2}}$ s
4.8×10^{-3}	1.81×10^4
2.0×10^{-2}	4.35×10^3
7.8×10^{-2}	1.11×10^3

5.3 Summary of Section 5

1   A combination of the differential and integration methods, or variations based on them, can be used to establish the form of a rate equation, to determine the order of reaction and to calculate the rate constant at a given temperature. The particular approach favoured by an investigator may depend on a number of factors (for example, the results of preliminary experiments, or chemical experience, though these are never infallible). A somewhat 'idealized' approach is summarized in Box 5.

2   The method of initial rates can be used to investigate reactions in which the products interfere in some way with the overall reaction. (It may, of course, also be used for more straightforward reactions.) For the method to be viable, a sensitive analytical technique is required to monitor the initial rates of formation of *product* species.

3   Integrated rate equations can be used to calculate expressions for the half-life of a reaction: this is a useful means of communicating the time-scale on which a reaction occurs. The half-life of a first-order reaction is independent of initial concentration, and this is a useful diagnostic test for such a reaction. The half-life of a second-order reaction involving only one reactant depends on the reciprocal of the initial concentration of the reactant.

> **Box 5**
>
> - Determine the stoichiometry.
>
> - Propose a rate equation. (This may depend on evidence from preliminary experiments that establish which species in the reaction mixture affect the rate of reaction.)
>
> - At constant temperature, measure appropriate kinetic reaction profiles; use isolation techniques if necessary.
>
> - Use *either* the differential method *or* the integration method, depending on circumstances, to determine the partial orders and, hence, the overall order of reaction. The advantages and disadvantages of these methods are:
>
> | | Differential method | Integration method |
> | --- | --- | --- |
> | advantages: | simple, and direct, graphical method for determining partial orders of reaction | experimental data are fitted directly to various integrated rate equations to find the best fit |
> | disadvantages: | instantaneous values of the rate of reaction are difficult to determine from kinetic reaction profiles | the reaction must be studied to more than 50% completion |
>
> - Once the rate equation has been determined, use an *appropriate integrated rate equation* to determine graphically a value for the rate constant.

6 THE EFFECT OF TEMPERATURE ON THE RATE OF A CHEMICAL REACTION

Ask a chemist what happens when the temperature of a reaction is increased by 10 K, and he or she will probably reply that the reaction rate is doubled; the answer is based on a well-known 'chemical rule of thumb'. On this basis, if a reaction were carried out in an ice-bath and then the same reaction repeated in a steam-bath, the rate of reaction would increase by a factor of 2^{10} or 1 024-fold. Evidently the effect of temperature can be very marked. Indeed, the 'rule' is often found to be conservative, and a 10 K increment causes more like a tripling of reaction rate; thus the increase for a rise of 100 K would be $3^{10} = 59\,049$, which is very large.

In this Section we examine an empirical description of the effect of temperature on reaction rate. In due course you will see that this description produces an equation that has important theoretical implications.

The variation of reaction rate with temperature is shown in a revealing manner by plotting the rate constant, k_R, against temperature. Such a plot was given in Section 2 (Figure 3) to emphasize the marked effect of temperature on the rate constant for the reaction between thiosulfate ion and 1-bromopropane. A similar plot is given in Figure 12 for the decomposition of dinitrogen pentoxide in the temperature range 300 K to 380 K.

The nature of the *increase* of rate constant with temperature in both cases suggests that there is an *exponential* relationship between these quantities. In fact, the following empirical expression adequately describes the behaviour:

$$k_R = A \exp(-B/T) \tag{79}$$

where A and B are *constants* for a particular reaction. (Note that the quantity within the exponential is negative: this ensures that the rate constant increases with increasing temperature providing B is positive, as it almost invariably is.)

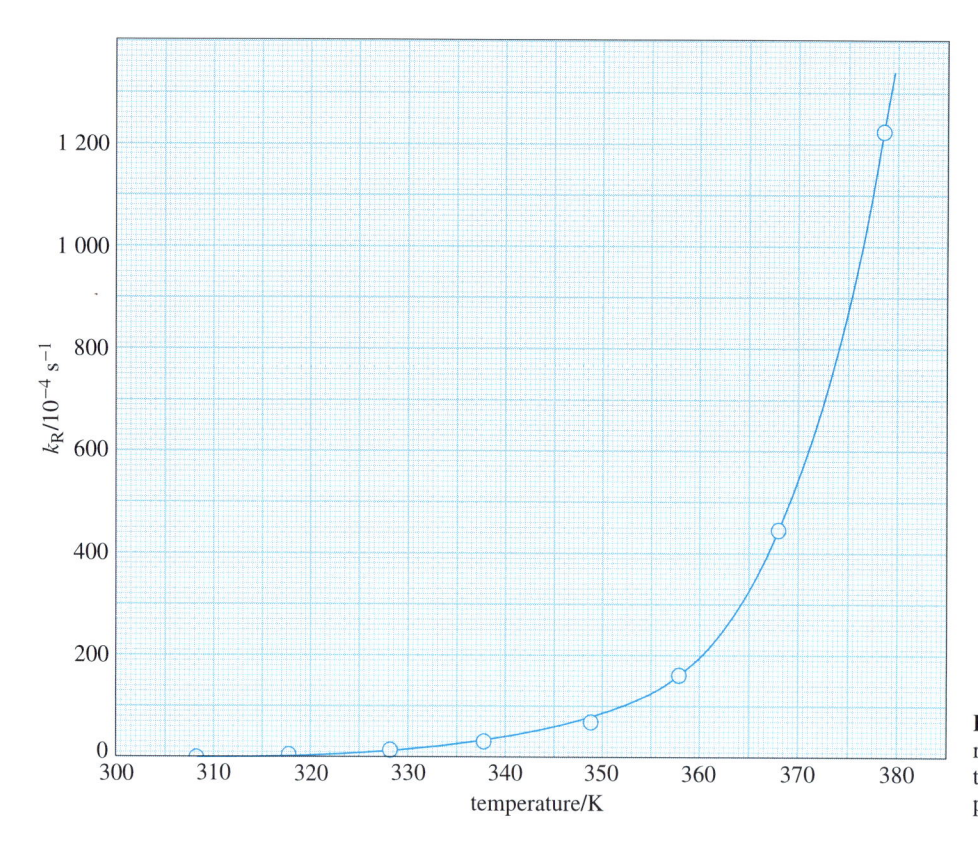

Figure 12 The variation of the rate constant with temperature for the decomposition of dinitrogen pentoxide.

The discovery of an exponential dependence of the rate constant on temperature for a whole variety of chemical reactions is attributed to the Swedish chemist Svante Arrhenius (Figure 13). Using semi-theoretical arguments, based on suggestions originally put forward by J. H. van't Hoff, he extended the form of equation 79 by equating the empirical constant B with the quantity E_a/R, where E_a has the dimensions of energy per mole. The more familiar form of the **Arrhenius equation** is thus:

$$k_R = A \exp\left(-\frac{E_a}{RT}\right) \tag{80}$$

The two parameters A and E_a are known collectively as the **Arrhenius parameters**. E_a is referred to as the **Arrhenius activation energy**, or more often just the *activation energy*. A is known by a variety of names: the **Arrhenius A-factor**, the pre-exponential factor, or the frequency factor. We shall refer to it as just the *A-factor*.

Figure 13 Svante Arrhenius (1859–1927). There is some controversy as to whether Arrhenius should be credited with the initial discovery of a general relationship between rate constant and temperature as in equation 79. J. H. van't Hoff, in a kinetic investigation carried out in 1884, proposed, but did not use, a relationship comparable with equation 79, and J. J. Hood of Glasgow asserted in 1885 that 'the temperature function is exponential', but did not suggest an equation. It was left to Arrhenius in 1889 to propose that equation 79 was generally applicable to all chemical reactions: he did not originate the relationship, but it is now justly named after him. Interestingly, Arrhenius was not primarily a kineticist. His doctoral thesis, concerned with electrochemistry, was not highly regarded by his examiners: however, he became Professor of Physics at Stockholm University in 1895, and in 1903 was awarded the third Nobel Prize for Chemistry – for formidable achievements in electrochemistry.

BLOCK 2 AN INTRODUCTION TO CHEMICAL KINETICS

■ What are the units of the *A*-factor?

▪ Since the exponential term is a pure number, these will be the *same* as the units for the rate constant. For instance, they will be s^{-1} for a first-order reaction.

The Arrhenius equation accounts very well for the temperature behaviour of a large number of reactions, and it is for this reason that it is widely used. It must be emphasized that the Arrhenius parameters are *experimental* quantities, although in due course you will learn that they can be given some physical meaning; indeed, you will see why the quantity E_a is termed an activation energy.

How would you determine the Arrhenius parameters from experimental measurements?

The answer is to write equation 80 in a linear form and then use a graphical procedure. This can be achieved by taking natural logarithms, when equation 80 becomes:

$$\ln k_R = \ln A - \frac{E_a}{RT} \tag{81}$$

By convention, *but no more than this*, it is common to take logarithms to the base ten, so recalling the relationship $\ln x = 2.303 \log x$, equation 81 becomes:

$$\log k_R = \log A - \frac{E_a}{2.303 RT} \tag{82}$$

Hence, provided A and E_a are indeed constants, a plot of $\log k_R$ versus reciprocal temperature ($1/T$) – which is called an **Arrhenius plot** – should be a straight line, from which the desired parameters can be found.

STUDY COMMENT You should now try SAQ 16 before moving on, because it provides 'hands-on' experience of using equation 82.

SAQ 16 Figure 14 (overleaf) shows an Arrhenius plot for the decomposition of N_2O_5. Use the information in the Figure to determine an activation energy for the decomposition reaction.

The determination of the *A*-factor is not as straightforward. According to equation 82, the straight line intercepts the $\log k_R$ axis (that is, when $1/T$ is zero) at a value of $\log A$. Examination of Figure 14 shows that the plot does not extend to this value of reciprocal temperature.

What happens if we choose the scale for the reciprocal temperature axis to extend to zero? Figure 15 shows an Arrhenius plot on an extended scale. It is clear from the Figure that the experimental points are now clustered together because of their relatively narrow temperature range. The determination of $\log A$ consequently involves a long, and usually unreliable, extrapolation if it is attempted by eye.

A better procedure* is to return to the Arrhenius plot in Figure 14: because the slope of the straight line in this plot is known (see SAQ 16), the *A*-factor is easily calculated from any point (not necessarily an experimental one) on the graph by using equation 82. Thus, for example, $\log (k_R/s^{-1}) = -2.84$ when 10^3 K/T = 3.0, or $1/T$ = 3.0×10^{-3} K^{-1}, so that

$-2.84 = \log A - (5.46 \times 10^3 \text{ K}) \times (3.0 \times 10^{-3} \text{ K}^{-1})$

$\log A = 13.54$

The *A*-factor has the same unit as k_R, so $A = 3.47 \times 10^{13}$ s^{-1}.

---

* Many modern calculators, and home computers with the appropriate software, provide a linear regression, or least squares, analysis for determining the 'best' slope and intercept of a straight line. Clearly, these can be used to advantage to find the appropriate values of E_a and A. However, it is always good practice to first plot a graph to make sure that any experimental data do 'fit' the Arrhenius equation, before carrying out the calculations.

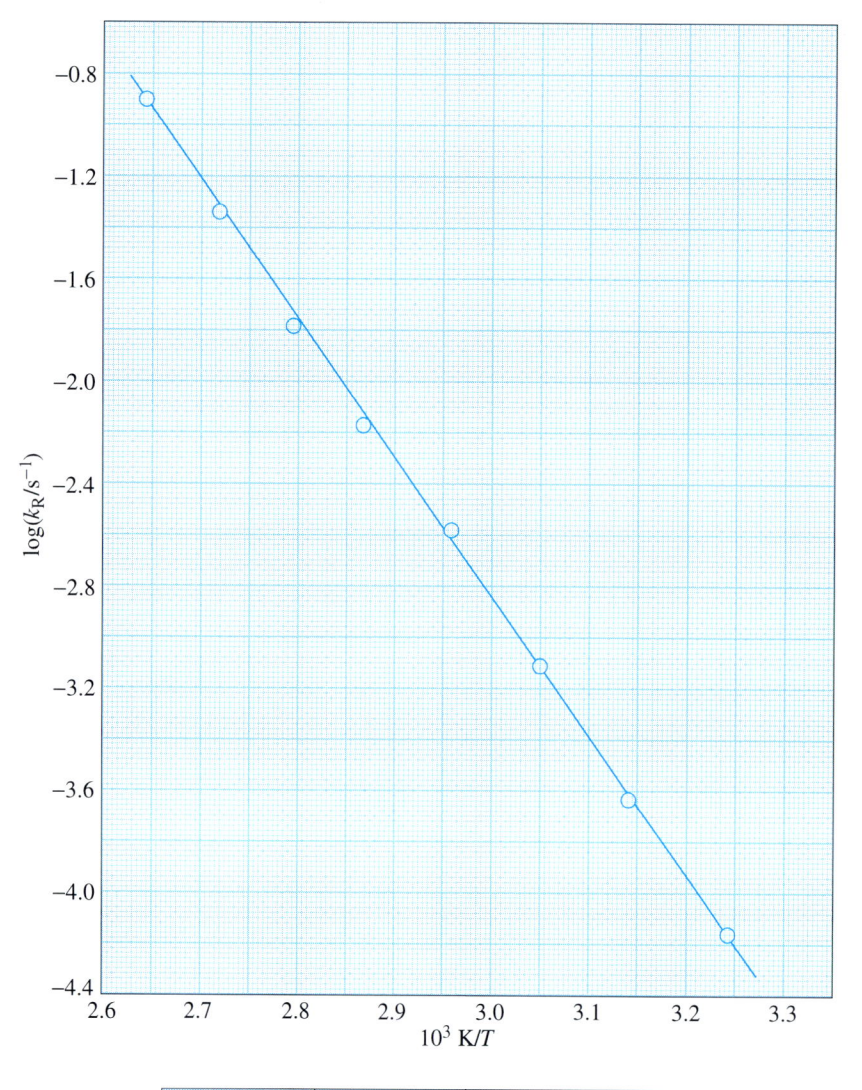

Figure 14 An Arrhenius plot for the decomposition of N_2O_5.

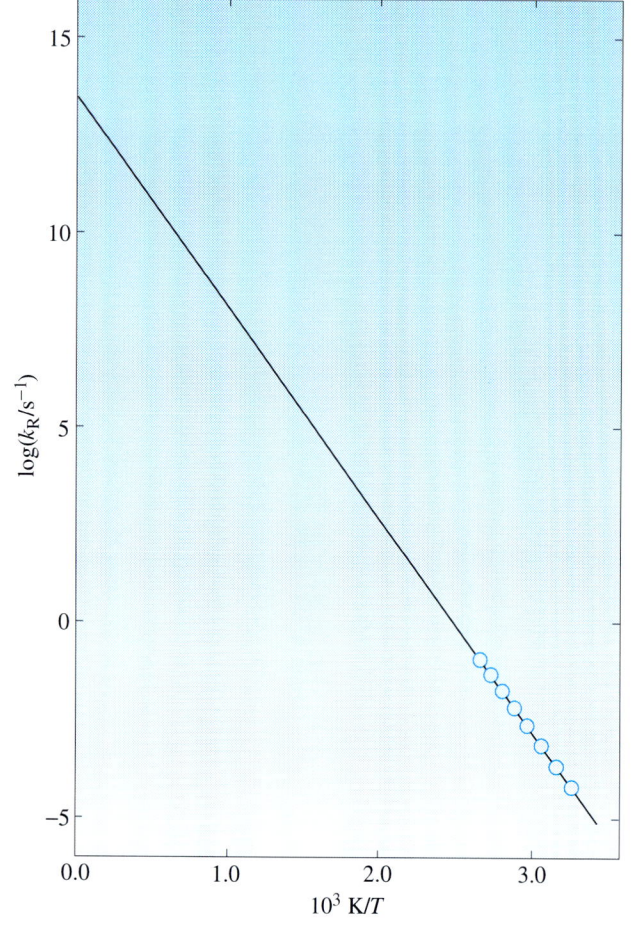

Figure 15 An Arrhenius plot for the decomposition of N_2O_5 on an extended scale.

Many chemical reactions can be studied over only a limited temperature range. The reason for this, apart from any changes in phase that the reaction mixture may experience, is experimental. Invariably, at high temperatures a reaction may be too fast to study by conventional methods, whereas at low temperatures it may become impracticably slow. These problems normally confine kinetic studies to a temperature range of 100 K or less. The Arrhenius plot in Figure 14 is thus typical of that encountered for many reactions.

It is worth re-emphasizing that the Arrhenius equation is remarkable in that it describes the temperature behaviour of a vast number of reactions. There are a few exceptions. Figure 16 shows in schematic form two examples of **non-Arrhenius behaviour** in terms of plots of the rate constant versus temperature.

You may be able to guess the situation represented by Figure 16a. It represents an explosion: the rate constant increases in a normal, that is exponential, manner until a critical ignition temperature is reached, at which point there is a sudden and violent increase in rate constant. Figure 16b depicts the behaviour sometimes observed in reactions catalysed by enzymes; at a certain temperature the enzyme 'decomposes' and the rate of reaction falls dramatically.

Finally, it is interesting to note that the Arrhenius equation has been found to describe the effects of temperature changes in some unusual situations. For instance, it applies to the chirping of tree-crickets (see Figure 17), the creeping of ants, and, allegedly, even psychological processes such as counting and forgetting.

SAQ 17 A keen naturalist enters a forest when the temperature is 285 K. Sometime later she notices that the rate of chirping of tree-crickets has doubled in value compared with when she first entered the forest. What increase in temperature has occurred? (Assume that the rate of chirping of tree-crickets obeys the equation, 'rate = constant × exp $(-E_a/RT)$', and use the value of E_a given in Figure 17.)

STUDY COMMENT   Do not miss out the following SAQ. It provides you with the opportunity to draw together some of the main points in this and earlier Sections.

SAQ 18 In alcoholic solution, iodomethane, CH_3I, reacts with ethoxide ion, $CH_3CH_2O^-$, according to the equation:

$$CH_3I + CH_3CH_2O^- = CH_3OCH_2CH_3 + I^- \tag{83}$$

Explain, *without giving experimental detail*, the steps you would take to determine the Arrhenius parameters for this reaction. [*Hint* Note very carefully that the *only* information given is the stoichiometry of the reaction.]

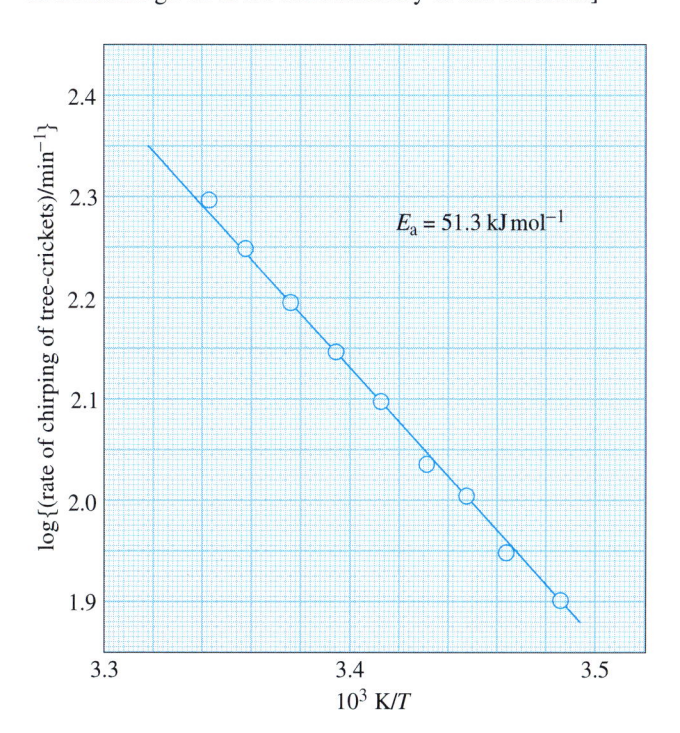

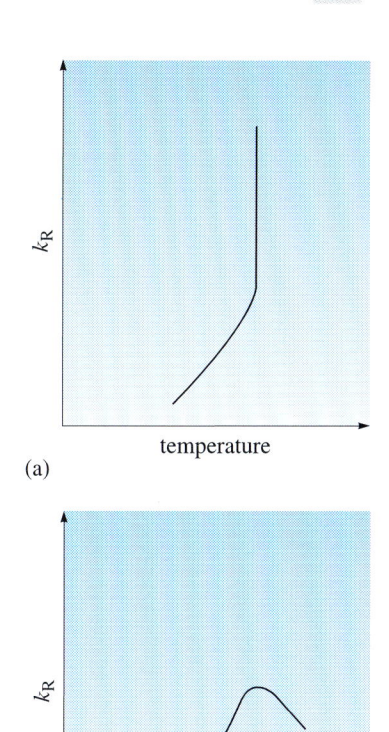

Figure 16 Non-Arrhenius behaviour: schematic plots of rate constant versus temperature.

Figure 17 An Arrhenius plot for the chirping of tree-crickets. (Source of data: K. J. Laidler, *J. Chem. Educ.*, 1972, vol. 49, p. 343)

$E_a = 51.3 \text{ kJ mol}^{-1}$

6.1 Summary of Section 6

1 The Arrhenius equation and its logarithmic forms are collected together in Box 6.

> **Box 6**
>
> $$k_R = A \exp\left(-\frac{E_a}{RT}\right) \quad (80)$$
>
> $$\ln k_R = \ln A - \frac{E_a}{RT} \quad (81)$$
>
> $$\log k_R = \log A - \frac{E_a}{2.303\,RT} \quad (82)$$

2 The Arrhenius equation is an empirical equation, which accounts very well for the temperature behaviour of a vast number of reactions.

3 The Arrhenius parameters are best determined graphically, as summarized in Box 7.

> **Box 7**
>
> Plot $\log k_R$ versus $1/T$ on a *suitable* scale.
>
> The slope of the resulting straight line is equal to $-E_a/2.303R$, so that E_a can be determined.
>
> Any point on the straight line may then be used to determine the *A*-factor, using either of the logarithmic equations in Box 6.

4 A few reactions show non-Arrhenius behaviour; for example, those involving explosions and, in some circumstances, those involving catalysis by an enzyme.

STUDY COMMENT This now completes our empirical description of the kinetics of chemical reactions, that is, the description of the effects of both concentration and temperature on the rate of a reaction. Some of the experimental techniques commonly used to investigate reaction rates are outlined in the next Section. Experimental techniques are also illustrated in several of the video sequences, and you will gain practical experience at the Residential School.

7 EXPERIMENTAL METHODS

The aim of most experiments in chemical kinetics is to monitor the concentration of one, or several, of the species in the reaction mixture as a function of time. A wide variety of techniques has been developed to achieve this aim. In this Section, we briefly outline just a small selection of the methods that are available.

7.1 Conventional techniques

In these techniques, the reaction is started by simply mixing the reactants together. In addition, it may also be necessary to bring the reaction mixture to the desired reaction temperature (particularly if only a single reactant is involved). Both the processes of mixing and heating (or cooling) take a finite time, typically of the order of one or two seconds. Thus conventional techniques can be applied with any accuracy only to reactions with half-lives of 10 s or longer; these can be described as relatively slow reactions.

As an example, consider the reaction we looked at in Section 2, the reaction between thiosulfate ion and 1-bromopropane:

$$S_2O_3^{2-} + C_3H_7Br = C_3H_7S_2O_3^- + Br^- \quad (5)$$

The reaction is carried out in an open flask: under these conditions it can be assumed to occur at *constant volume* because the volume change on reaction will be very small. Since (as you learnt in Section 6) the rate of reaction can be markedly temperature dependent, it is essential that the reaction vessel be maintained at *constant temperature* by means of a thermostat. (To determine a rate constant to an accuracy of ±2.5% at 298 K for a reaction with an activation energy of about 100 kJ mol^{-1}, the temperature must be controlled to within ±0.2 K.)

One way to follow the reaction is to use a volumetric method; the concentration of thiosulfate ion can be determined at different times in the reaction by titration with a standard iodine solution (the analytical details can be found in standard textbooks). The efficient way to do this is to withdraw small samples from the reaction mixture at selected times. Of course, for this strategy to be successful, the time taken for sampling and titration must be very short compared with the time during which significant changes in the composition of the reaction mixture occur. To avoid this time limitation, it is best to **quench**, or stop, the reaction as soon as the sample is taken. This is achieved by rapidly cooling the sample to a temperature at which the reaction rate is very small: an ice-bath is sufficient in this case. (An alternative would be to remove the thiosulfate ion completely, by adding the sample to an excess of iodine solution, and then to determine the amount of excess iodine volumetrically.) Once the sample is quenched it may be analysed at leisure.

This example illustrates the steps to be taken in a standard experiment in chemical kinetics. These are: arrange that the reaction occurs at constant temperature, start the reaction by mixing the reagents, sample the reaction mixture at various times, quench the individual samples, and analyse them using a suitable technique. The process of sampling reduces the *mass* of the reaction mixture, but it does not change the concentrations, and so it cannot affect the rate of reaction – a vital prerequisite of any analytical technique. Sampling is, however, a rather time-consuming technique and it is preferable, if possible, to monitor the concentration of a reactant or product *directly* as the reaction proceeds: this requires a **physical method of analysis.**

Physical methods rely on the measurement of a property of the reaction mixture, for example pressure, absorbance, or electrical conductivity, which is related, preferably in a linear manner, to the concentration of one (and sometimes of several) of the species involved in the reaction. The limitation is that the time required to make a measurement (that is the *response time* of the technique) *must be short* compared with the time in which significant changes in concentration occur. Below we outline, albeit briefly, just a few of the more common methods.

7.1.1 Pressure measurement

If there is a change in the total number of molecules during a gas-phase reaction, and if the reaction is studied in a constant-volume system, then measurement of the *total pressure* as a function of time can be used to follow the reaction. For example, consider the decomposition of dimethyl ether, CH_3OCH_3, in the gas phase:

$$CH_3OCH_3(g) = CH_4(g) + H_2(g) + CO(g) \tag{84}$$

- Write an expression for the total pressure, p_{tot}, at any time in the reaction in terms of the partial pressures of the constituent gases.

- If we assume that the individual gases and the mixture behave ideally, then, according to the discussion in Block 1, Dalton's law of partial pressures tells us that:

$$p_{tot} = p(CH_3OCH_3) + p(CH_4) + p(H_2) + p(CO) \tag{85}$$

At the start of the reaction $p(CH_3OCH_3) = p_0$, where p_0 is the initial pressure, and $p(CH_4) = p(H_2) = p(CO) = 0$. At some later time, the stoichiometric equation tells us that the *increase* in partial pressure of say CH_4, must equal the *decrease* in partial pressure of CH_3OCH_3; in fact,

$$p(CH_4) = p(H_2) = p(CO) = p_0 - p(CH_3OCH_3) \tag{86}$$

- What is the relationship between the partial pressure of dimethyl ether at a given time in the reaction and the total pressure at this time?

- If we substitute equation 86 into equation 85:

$$p_{tot} = p(CH_3OCH_3) + 3\{p_0 - p(CH_3OCH_3)\}$$
$$= 3p_0 - 2p(CH_3OCH_3) \tag{87}$$

so that, on rearranging,

$$p(CH_3OCH_3) = \frac{3p_0 - p_{tot}}{2} \tag{88}$$

This calculation illustrates how the partial pressure of a reactant, dimethyl ether in this case, can be related to its initial pressure and the total pressure at a given time in the reaction. Total pressure measurements can thus be used to establish a kinetic reaction profile, from which the rate equation can be established.

Note very carefully that it is essential that the stoichiometry of the reaction is adequately represented by the chemical equation on which the calculations are based; that is, the contribution of any concurrent or side reaction must be negligible.

The experimental arrangement for studying the kinetics of the gas-phase decomposition of 1,2-dichloroethane to give chloroethene (vinyl chloride) is demonstrated in video band 1. The sequence also explains how **gas chromatography** can be used to investigate this reaction.

STUDY COMMENT   The following SAQ introduces an alternative method of carrying out 'partial pressure calculations' for gas-phase chemical reactions. You should make sure that you attempt to answer it.

SAQ 19 For a gas-phase reaction, occurring in a constant volume system at a particular temperature, show that a 'modified reaction variable', x', can be defined as follows

$$x' = \frac{p(Y) - p_0(Y)}{\nu_Y} \tag{89}$$

The quantity x' is related to the reaction variable, x, (Section 3.1) by $x' = xRT$, and it is assumed that the individual gases, as well as the reaction mixture, behave ideally.

Using this 'modified reaction variable', show that the total pressure at any time in the thermal decomposition of N_2O_5

$$2N_2O_5(g) = 4NO_2(g) + O_2(g) \qquad (8)$$

can be expressed as

$$p_{tot} = p_0(N_2O_5) + 3x' \qquad (90)$$

and, hence, confirm that the partial pressure of N_2O_5 at any time in the reaction mixture is

$$p(N_2O_5) = \tfrac{1}{3}\{5p_0(N_2O_5) - 2p_{tot}\} \qquad (91)$$

where $p_0(N_2O_5)$ represents the initial pressure of N_2O_5.

7.1.2 Spectroscopic measurement

Spectroscopic techniques are often used in making kinetic measurements. At the simplest level, the principle of their application is quite straightforward: the intensity of some spectral line characteristic of a reactant or product species is measured as a function of time, the conversion of intensity into concentration being achieved by prior calibration of the spectroscopic equipment. The spectral ranges you may be familiar with are those involving visible and ultraviolet absorption, or those involving infrared absorption.

Visible and ultraviolet spectrophotometry is a very common means of measuring concentration. Each substance has its own characteristic absorption spectrum (its origin lies in transitions between different electronic states), which is essentially a measure of the absorption of light as a function of wavelength. The **Beer–Lambert law** relates the amount of radiation absorbed *at a given wavelength* to the concentration of the absorbing species:

$$\log\left(\frac{I_{in}}{I_{out}}\right) = \epsilon c l = A \qquad (92)$$

I_{in} is the intensity of the incident monochromatic light, I_{out} is the transmitted intensity, l is the path length, c is the concentration, and ϵ is the *molar absorption coefficient*. The dimensionless quantity A is called the *absorbance*, the quantity measured by the spectrophotometer. As an example, Figure 18 shows a plot of absorbance versus wavelength for bromine dissolved in carbon tetrachloride.

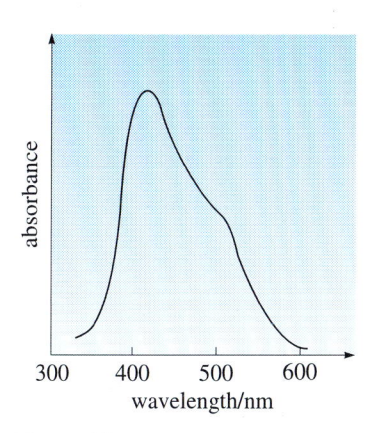

Figure 18 Absorption spectrum of bromine dissolved in carbon tetrachloride.

Equation 92 shows that absorbance is directly proportional to concentration. Hence, if a wavelength can be chosen such that the molar absorption coefficient is large for either a reactant or a product species, a reaction can be followed by simply monitoring the absorption of light by this substance*. For instance, the conversion of the hypochlorite ion into the hypoiodite ion, which we examined in Section 5.1.1:

$$I^-(aq) + ClO^-(aq) = IO^-(aq) + Cl^-(aq) \qquad (1)$$

can be followed spectrophotometrically, because the hypoiodite ion absorbs light strongly at 400 nm, whereas the other species in the reaction mixture do not.

7.1.3 Electrochemical measurement

You will learn about electrochemical aspects of chemical reactions in Blocks 7 and 8. For the present, you should simply be aware that if a reaction occurring in solution involves a change in the number and type of ions present, then this affects the *electrical conductivity* of the solution. Thus, measurements of conductivity can be used to monitor the progress of a reaction.

---

* In more general terms, it is important to note that the *total* absorbance of a reaction mixture is equal to the sum of the absorbances of the individual species present. Given the stoichiometry of a reaction, it is thus possible (as with total pressure measurements) to relate total absorbance measurements to the concentrations of individual components in the mixture.

7.2 Flow methods

The term 'fast reaction' can be used to refer to a reaction that is fast relative to the time required for mixing and observation by conventional methods. However, the term is relative; for example, very modern laser techniques are capable of probing chemical reactions on the femtosecond (10^{-15} s) time-scale. In general, an impressive array of specialized, and very elegant, techniques has been developed for studying fast reactions. These include relaxation methods, such as temperature jump and pressure jump, photolysis methods, methods based on nuclear magnetic resonance and electron spin resonance spectroscopies, pulsed laser techniques, and many more. Here, we have space to consider only one type, based on **flow methods**.

Flow methods can be traced back to the pioneering work of H. Hartridge and F. J. W. Roughton (1923), who were the first to design special mixing chambers that allowed two reactant solutions to be mixed within about 10^{-3} s. In one variation, the **continuous flow method**, two reagents are rapidly mixed in an optimally designed mixing chamber and then the thoroughly mixed solutions flow through an outlet tube, 'reacting as they go'; a schematic diagram of a typical experimental arrangement is shown in Figure 19. Without going into detail, it turns out that a *steady state* is eventually set up along the outlet tube.

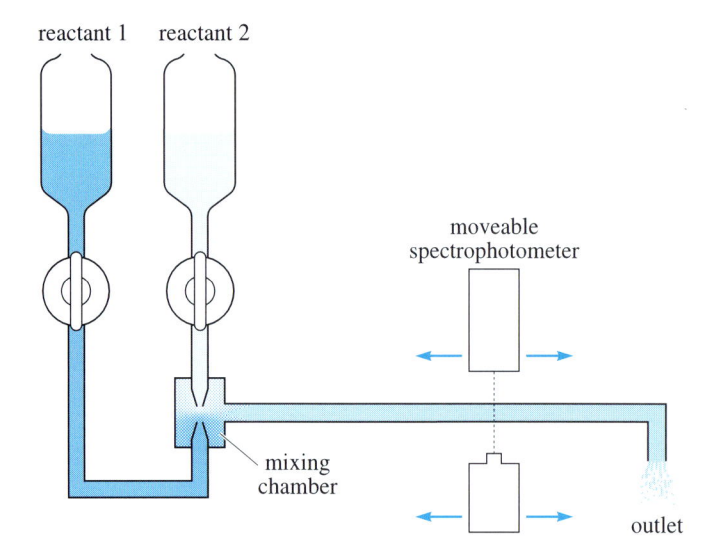

Figure 19 Schematic diagram of a continuous flow apparatus.

The concentration of a reactant, at different *distances* along the tube, measured by a suitable spectroscopic technique, in effect corresponds to observing the reaction mixture at different *times* from the commencement of mixing. Once the steady state is established, the experimental measurement of concentration can be made at leisure, *providing* that sufficiently large volumes of solution are available. The method is suitable for reactions occurring on a time-scale of milliseconds.

The **stopped-flow method** is preferable when only limited amounts of reagents can be used, and it is employed quite often in investigations of reactions catalysed by enzymes. You will carry out a kinetic investigation using this method at the Residential School.

7.3 Summary of Section 7

1   Experimental methods for studying the kinetics of reactions can be divided into conventional and 'fast reaction' techniques.

2   The time taken for mixing and observation in conventional methods limits their application to reactions with half-lives of 10 s and longer.

3   If a sampling technique is used, it is preferable to quench the sample immediately it is taken, so that it may then be analysed at leisure.

4   Physical methods allow the concentration of a reactant or product to be monitored directly as a reaction progresses. Properties such as pressure, absorbance and electrical conductivity are commonly measured because these are related to concentration in a linear manner. The response time of the technique used must be short compared with the time in which significant changes in the composition of the reaction mixture occur.

5   Flow methods are suitable for the investigation of reactions that occur on the millisecond time-scale.

SAQ 20   Solutions of potassium trioxalatocobaltate(III) are rapidly reduced by iron(II) sulfate in dilute solution. The overall reaction can be represented by the following equation:

$$Fe^{2+} + (Co(C_2O_4)_3)^{3-} = Fe^{3+} + Co^{2+} + 3C_2O_4^{2-} \tag{93}$$

where $(Co(C_2O_4)_3)^{3-}$ is the trioxalatocobaltate(III) ion. The experimental rate equation is:

$$J = k_R[Fe^{2+}][(Co(C_2O_4)_3)^{3-}] \tag{94}$$

and at 318.5 K the value of k_R is found to be 9.8×10^2 dm^3 mol^{-1} s^{-1}.

What is the half-life of the reaction at 318.5 K when the initial concentrations of Fe^{2+} and $(Co(C_2O_4)_3)^{3-}$ are both equal to 1.0×10^{-3} mol dm^{-3}? The reaction was, in fact, investigated using conventional methods at 318.5 K. Can you suggest how this was achieved?

8 REACTIONS BETWEEN MOLECULES:
ELEMENTARY REACTIONS

For the remainder of this Block we turn our attention to the description of theoretical models of chemical reactions. Our aim will be to examine just how chemical reactions occur between individual molecular and atomic species.

At the microscopic level, the basic act of chemical change is a process that invariably involves just one or two chemical species – atoms, molecules, ions, radicals or electrons – which are transformed into one or more new species. These *individual* events are called **elementary processes**: they occur in a single step and do not involve the formation of any reaction intermediate. An overall chemical reaction may involve many repetitions of just a single molecular process – in which case it is again referred to as an elementary, or sometimes *simple*, reaction – or it may involve many repetitions of a series of such processes; that is, the reaction has a multi-step mechanism and is sometimes referred to as a *composite* reaction. This type of reaction will be studied in more detail in Block 3.

Elementary reactions can be divided, although rather artificially, into collisional and decay processes. A collisional process, as its name implies, involves two (very rarely three) chemical species undergoing a collision, which results in a chemical transformation. A common example, which we shall examine in due course, involves the transfer of a single atom, such as in the gas-phase reaction:

$$F\cdot + H_2 \longrightarrow HF + H\cdot \tag{95}$$

The manner in which the fluorine atom is represented takes into account the fact that it has an 'unpaired electron': this is signified by the dot.

Decay processes involve a single species that changes to a different form. For example, a molecule may dissociate:

$$Br_2 \longrightarrow Br\cdot + Br\cdot \tag{96}$$

or it may rearrange into a different isomeric form:

$$CH_3NC \longrightarrow CH_3CN \tag{97}$$

It is important to appreciate that decay processes are more complex than chemical equations such as equations 96 and 97 imply, because before a molecule may dissociate or rearrange it must first become energized or activated: it is *collisions* between reactant molecules themselves (or with other species that may be present) that provide this energy. Thus, the definition of an elementary reaction as one taking place in a single step is *independent* of the way that the reactants gain the necessary energy to react.

Notice the nomenclature we have used to depict the elementary reactions in equations 95, 96 and 97. So far we have represented chemical reactions using the equality sign (=); this is the agreed sign for a balanced chemical equation. The arrow sign (⟶) is used to indicate a reaction that is known or postulated to be elementary; *it implies that the reaction does proceed*. (In some literature you may find this symbol used to indicate a composite reaction; indeed, we adopt this notation later in the Course, although the context in which it is used will always be made clear.) The symbol ⇌ is used as a shorthand notation to indicate a reaction for which both the forward and reverse elementary steps are important: you may recall that this symbol was used in the reaction mechanism given as an example in Section 1.

A common method of classifying elementary reactions is in terms of the *number* of reactant molecules that take part: this is called the **molecularity**. Elementary reactions are thus *unimolecular, bimolecular* or *termolecular* according to whether one, two or three reactant molecules are involved, respectively. There are very few examples of termolecular reactions, and elementary reactions between four or more chemical species are unknown: it would be highly improbable that four reactant species could all collide and undergo chemical transformations at the same instant in either a gas-phase or a solution reaction. It should be noted that elementary reactions involving a single species (such as reactions 96 and 97) are classified as unimolecular even though the molecules must first become activated by collisions 'with other bodies' before they can dissociate or rearrange.

> It is very important to make a clear distinction between classifying an *overall* reaction in terms of its order, and classifying an *elementary* reaction in terms of its molecularity. Molecularity is a theoretical concept, which refers to a postulated elementary reaction, and it has a value that is necessarily a small integer (1, 2 or, very rarely, 3). Order is an empirical concept, which is determined experimentally, and may have a positive, negative, fractional or zero value.

■ If the order of a reaction is known, can the molecularity be stated?

▨ No. The only circumstances in which the molecularity of a reaction can be stated is if it is known, or postulated, that the reaction occurs in a single step. *This information cannot be deduced from a knowledge of the order of reaction alone.*

For bimolecular reactions, the rate of reaction will depend on the concentrations of the two reactant molecules (we discuss this in more detail in Sections 9 and 10) and so it is possible to write down the rate equation directly; thus, for the reaction described by equation 95:

$$J = -\frac{d[F\cdot]}{dt} = \frac{d[HF]}{dt} = k_R[F\cdot][H_2] \tag{98}$$

For unimolecular reactions, under most experimental conditions, the rate of reaction is directly proportional to the concentration of the single reactant; thus, for the reaction described by equation 97:

$$J = -\frac{d[CH_3NC]}{dt} = \frac{d[CH_3CN]}{dt} = k_R[CH_3NC] \qquad (99)$$

Equations 98 and 99 demonstrate, in general, that:

> For an elementary reaction, the overall order of reaction and the molecularity are the same, so the rate equation can be written down by inspecting the stoichiometry.

It should be noted that for atom recombination reactions such as

$$Br\cdot + Br\cdot \longrightarrow Br_2 \qquad (100)$$

the situation is a little more complex. This is because if a Br_2 molecule is to be formed, the pair of Br atoms must collide with a 'third body' while they are close together, so that this body can remove a significant part of the kinetic energy that the pair possess. If this did not happen they would simply bounce apart, without the formation of a chemical bond. However, the rate equation can be, and usually is, written as:

$$J = -\frac{1}{2}\frac{d[Br\cdot]}{dt} = \frac{d[Br_2]}{dt} = k_R[Br\cdot]^2 \qquad (101)$$

but it should be remembered that the value of the 'rate constant', k_R, depends on the nature *and* concentration of the 'third body'. The 'third body' is often denoted by the symbol M.

Finally, from a practical viewpoint, there is a very important restriction on the statement that a particular reaction is elementary. This is best illustrated by an example. For many years – from the turn of the century to the 1960s – the gas-phase reaction between hydrogen and iodine:

$$H_2(g) + I_2(g) = 2HI(g) \qquad (102)$$

was taken as the textbook example of how molecules could exchange partners in a single elementary step. It thus came as a shock to many chemists when in the early 1960s, meticulous experimental work by J. H. Sullivan revealed that very little of the hydrogen iodide was produced by a bimolecular reaction. The evidence pointed to the production of hydrogen iodide by not one, but two, distinct mechanisms, each starting with the dissociation of iodine into its atoms. It is salutory to reflect that such an apparently 'well-understood' reaction should, after more than sixty years, be shown to be so complex.

The moral of this cautionary tale is that the statement that a specific reaction is elementary is *always* subject to the limitation that further evidence may eventually come to light showing that it is composite.

8.1 Summary of Section 8

1   An elementary reaction is one that takes place in a single step which does not involve the formation of any reaction intermediate. This definition is independent of the way the reactants gain the necessary energy to react.

2   The molecularity of an elementary reaction is a method of classifying a reaction in terms of the number of reactant species that take part.

3   Elementary reactions may be unimolecular, bimolecular or, very rarely, termolecular.

4   Molecularity is a theoretical concept and must not be confused with reaction order, which is an experimentally determined quantity.

5   In general, unless stated otherwise (when usually a further mechanism involving the physical process of energy activation would be given), the rate equation for an elementary reaction can be written down by inspecting the stoichiometry.

6   A reaction may be taken or postulated to be elementary only until evidence to the contrary is discovered.

9 COLLISION THEORY
OF CHEMICAL REACTIONS

Our discussion in the previous Section suggested a very basic model for an elementary reaction: chemical species must meet, or collide, before any chemical transformation can occur. **Collision theory** simply formalizes and quantifies this intuitive idea. In this Section we examine how the theory applies to both gas-phase and solution-phase *bimolecular* reactions.

The best procedure for testing the theory is to use it to predict the value of bimolecular, second-order rate constants. But this immediately raises a problem. The experimental rate constant reflects an average over a multitude of individual molecular events, so as a consequence a *molecular* theory must also be averaged in a suitable way. This is most 'easily' accomplished for gas-phase reactions, because the description of the behaviour of molecules in gases is well developed. In solution the problems are more formidable; the theory of the liquid state is, at best, rudimentary and, in addition, reactant species are always surrounded by solvent molecules whose exact influence is difficult to disentangle. Thus, we first look at collision theory for gas-phase reactions, and then see what modifications are necessary to make it useful in solution kinetics. However, it is well to keep in perspective that although gas-phase reactions may be easier to deal with theoretically, it is in solution that a great deal of chemistry occurs and where the bulk of chemical investigation is carried out.

9.1 The gas phase

The collision theory of gas-phase reactions was first formulated around 1920. It proposes that the rate of a bimolecular reaction depends on two factors: first, how often molecules* collide (that is, the **collision frequency**), and second, the fraction of these collisions that are effective in bringing about chemical transformation. Thus, in schematic form, the rate of a bimolecular gas-phase reaction can be written as:

> rate = (collision frequency) × (fraction of collisions which lead to chemical transformation) (103)

Both of the terms in parentheses can be given mathematical form; however, since we are more interested in the predictive powers of the theory rather than its detailed development, we shall outline only the key features of the analysis.

9.1.1 Calculating a collision frequency

The first problem is to decide how to define a collision. The answer depends very much on the properties we choose to give the molecules. Certainly, the more complex (in effect this means real) we make a model of a collision, the more difficult, and ultimately intractable, will become the mathematics. The simplest model we can select makes the following assumptions:

- Molecules are regarded as being spherical with definite diameters. This effectively abandons the idea of any internal structure, and so denies the molecule any vibrational motion.

- All attractive forces between molecules are ignored.

- A very large repulsive force exists between molecules when they touch; that is, they are *hard* (like snooker balls).

---

* We use the term 'molecules' in a general sense here to represent atoms, molecules, 'fragments of molecules' or ions. In discussing theories of reactions we shall adopt this shorthand on several occasions – but it will always be clear in which context the term is used.

The assumptions of this **hard-sphere model** are drastic. Nevertheless, they have the benefit that they make collisions easy to define. As you will see, they also allow construction of a theory that gives broad insight into factors that are important in determining reaction rates. As is sometimes the case in science, a simple approach to a complex problem can be rewarding.

To focus our thoughts, consider an elementary reaction between two different molecules A and B:

$$A + B \longrightarrow \text{products} \tag{104}$$

(F· and H_2, say, as in equation 95), occurring in a system of constant volume V. Initially, there are N_A* molecules of A and N_B molecules of B; that is, there are N_A/V molecules of A per unit volume and N_B/V molecules of B per unit volume. (It is well to be clear from the outset that in any calculation using the SI system of units, volume must be measured in m^3, so *unit* volume is $1\,m^3$.) Figure 20 shows one pair of hard-sphere molecules A and B approaching, colliding and rebounding, a picture not too unlike that seen when two snooker balls are bounced off one another. The Figure illustrates how the collision between A and B is defined: the molecules collide when the distance separating their *centres* is d, where

$$d = r_A + r_B \tag{105}$$

and r_A and r_B are the radii of molecules A and B, respectively. The quantity d is called the **collision diameter**.

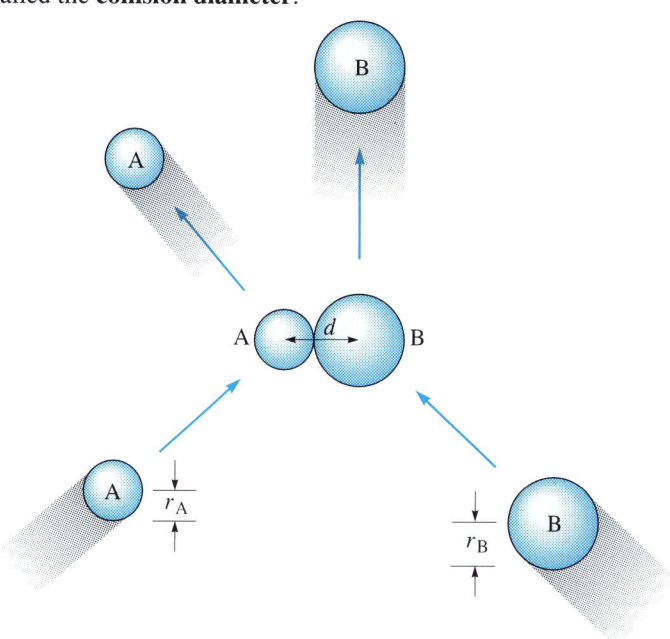

Figure 20 A schematic representation of a collision between hard-sphere molecules A and B.

So much for the definition of a collision: but how can we calculate the frequency with which such collisions occur? This is a fairly complex calculation, but insight can be gained, and important parameters introduced, by first considering a simplified situation. Suppose we have a mixture of gases A and B. Consider the B molecules to be stationary. In this picture the A molecules move in a 'sea' of randomly positioned B molecules. One such A molecule is shown in Figure 21, which also indicates how to tackle a quantitative calculation.

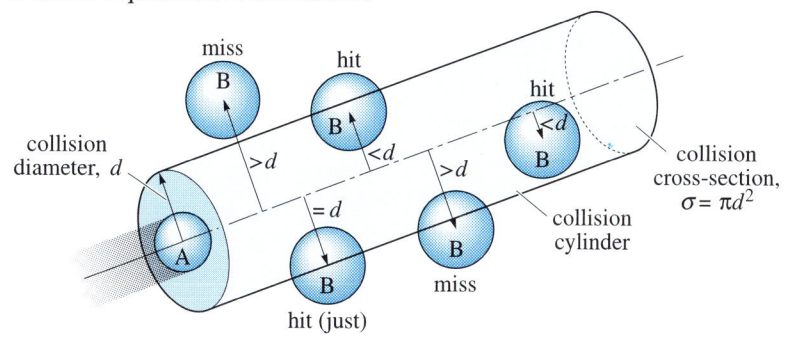

Figure 21 A schematic representation of the motion of a single A molecule through a gas of stationary B molecules: the collision cylinder and collision cross-section are indicated. Note that the diameter of the collision cylinder is $2d$.

* This is also the symbol for the Avogadro constant. To avoid confusion, in *this* Block we shall use the symbol L for the Avogadro constant.

This A molecule can be thought of as 'sweeping out' a **collision cylinder** of cross-sectional area, or, as it is better known, **collision cross-section**, σ, equal to πd^2. Of course, on collision, the direction of motion of the A molecule will change (see, for example, Figure 20), and so the direction of the collision cylinder will also change. However, because the B molecules are randomly positioned, we can, for illustrative purposes, join all the individual cylinders together to form one single cylinder.

- ■ If the A molecule has speed u_A, what will be the volume of the collision cylinder that it sweeps out in a time t?

- ▪ In a time t, the A molecule will travel a distance $u_A t$, and thus it will sweep out a collision cylinder of volume $\sigma u_A t$ (where $\sigma = \pi d^2$).

- ■ What is the *frequency* of collisions with B molecules? That is, how many collisions will occur in unit time?

- ▪ The number of B molecules with their centres lying within the collision cylinder is equal to the volume of the cylinder times the number of B molecules per unit volume. Hence, the number of collisions with B molecules in a time t is $(\sigma u_A t) \times (N_B/V)$, and the frequency (collisions per unit time) is $\sigma u_A N_B/V$.

But of course, there is more than one A molecule (there are N_A/V per unit volume), and so the *total* number of A-B collisions per unit time per unit volume is $\sigma u_A N_A N_B/V^2$.

To make the calculation more realistic, we must recognize that:

- Both molecules A and B will be in motion, and so the *relative* speed of A and B with respect to one another, denoted by u_{AB}, will be important. (The relative speed is the speed an 'observer' on molecule A would attribute to molecule B, or vice versa.)

- Both molecules A and B will have a characteristic *distribution* of molecular speeds. For a gas at a given temperature, the distribution of molecular speeds is accurately described by the Maxwell distribution. For example, Figure 22 shows the distribution of molecular speeds (plotted as the percentage of molecules with a particular speed versus that speed) for nitrogen gas at three temperatures. (The significance of the temperature variation will become apparent shortly.)

A common, and very useful, approximation is to consider just the *average collision properties* of the mixture. This is easily achieved: the calculations based on Figure 21 are repeated, *but* with the speed of molecule A (that is u_A) replaced by an *average*

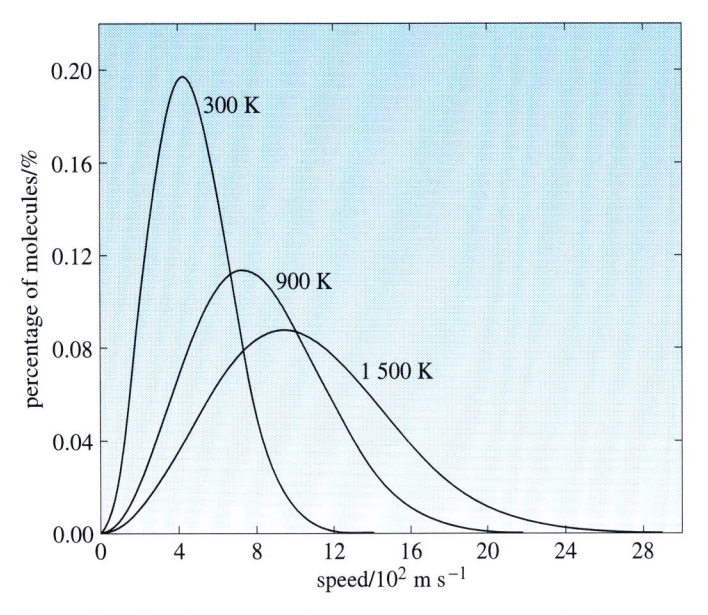

Figure 22 The distribution of speeds of nitrogen molecules in the gas phase at three different temperatures.

relative speed $\bar{u}_{AB}$. Thus, the total number of collisions between A and B molecules per unit time per unit volume, that is, the quantity we shall call the **total collision frequency per unit volume**, Z_{AB}, is then given by:

$$Z_{AB} = \sigma \bar{u}_{AB} \frac{N_A N_B}{V^2} \tag{106}$$

An expression for the average relative speed, $\bar{u}_{AB}$, can be derived on the basis that the relative speeds have a Maxwell type of distribution; without giving proof, we ask you to accept that:

$$\bar{u}_{AB} = \left(\frac{8kT}{\pi \mu}\right)^{1/2} \tag{107}$$

where k is the Boltzmann constant*, whose value is $1.380\,662 \times 10^{-23}\,\text{J K}^{-1}$, T is the temperature, and μ is the **reduced mass**. The latter quantity is defined as:

$$\frac{1}{\mu} = \frac{1}{m_A} + \frac{1}{m_B}, \text{ or } \mu = \frac{m_A m_B}{m_A + m_B} \tag{108}$$

where m_A and m_B are the masses of molecules A and B, respectively (expressed in kg).

SAQ 21 Often in calculations involving collision frequencies, the partial pressure of a gas (expressed in Pa) is given rather than its concentration in units of molecules per cubic metre. If a gas has a partial pressure of 10^5 Pa at 400 K, what is the concentration expressed in molecules m^{-3}?

SAQ 22 Determine the total collision frequency per unit volume, Z_{AB}, between ethene, C_2H_4, and buta-1,3-diene, $CH_2=CH-CH=CH_2$, in a mixture of the two gases in a sealed vessel at 400 K. Use a partial pressure of 10^5 Pa for each gas and assume the hard-sphere collision diameter is $d = 500$ pm (1 pm = 10^{-12} m). You may assume the gases behave ideally. [*Hint* Be very careful with units, and approach the calculation in stages.]

9.1.2 Reactive collisions

The total collision frequency for a bimolecular reaction must represent an *upper* limit to the rate of reaction: obviously, molecules cannot react faster than they collide. We emphasize it represents an *upper* limit because, for most reactions, only a fraction of collisions actually result in chemical transformation. This is dramatically illustrated by the gas-phase reaction between ethene, C_2H_4, and buta-1,3-diene, $CH_2=CH-CH=CH_2$, to give cyclohexene, C_6H_{10}:

$$C_2H_4(g) + CH_2=CH-CH=CH_2(g) = C_6H_{10}(g) \tag{109}$$

At 400 K the reaction is thermodynamically favourable, yet under the conditions described in SAQ 22 the reaction half-life is of the order of 37 *years*, even though the reactant molecules experience some 10^{35} collisions per second in a volume of 1 m^3.

■ Increasing the temperature to 500 K lowers the half-life substantially to about 13 days (which is still a long time). Can this be explained by the effect of temperature on the total collision frequency?

▪ No. As you can see from equations 106 and 107, the total collision frequency depends only on the square root of temperature.

Clearly, the second factor in our schematic expression (that is, equation 103) for the rate of a bimolecular reaction is very important.

---

* The symbol k is also used to represent a rate constant, but it is always given a subscript when used in that context.

The reason that only a fraction of collisions lead to reaction is that, in general, there is an **energy barrier to reaction**, an idea first introduced by Arrhenius to justify the form of the equation now named after him. To react, molecules must surmount this barrier by colliding with sufficient energy. So how do we incorporate the concept of an energy barrier into our hard-sphere collision theory?

The simplest assumption is that two potentially reactive molecules will react only when their relative translational energy is greater than some minimum or **threshold energy**, ε_0, or, in other words, they have a relative speed in excess of some minimum value. But there is a slight flaw in the model. Let us return to the snooker analogy: the results of two equally powerful shots, one resulting in a head-on collision but the other only a glancing collision, are very different. A better assumption would be that the two molecules will react only if they have sufficient kinetic energy *as measured along the line joining their two centres at the moment of impact*: energy 'perpendicular to this line' will be of no use in overcoming the energy barrier. For a Maxwell type of distribution of relative molecular speeds, it turns out (again, we ask you to accept this) that the fraction of *all* molecules having an energy greater than the threshold 'line-of-centres' energy is directly proportional to:

$$\exp(-\varepsilon_0/kT) = \exp(-E_0/RT)$$

where k is the Boltzmann constant, $E_0 = L\varepsilon_0$, and L is the Avogadro constant (equal to 6.022×10^{23} mol^{-1}).

The exponential dependence means that the effect of changing the temperature will be most marked. We can appreciate this to some extent by returning to Figure 22 (even though it is concerned with the distribution of molecular speeds in a single, unreactive gas).

■ What happens to the distribution of molecular speeds as the temperature is raised?

■ As the temperature is raised, the fraction of slow-moving (low-energy) molecules decreases, whereas the fraction of rapidly moving (high-energy) molecules increases. As the Figure shows, the change at high speed is most marked.

By analogy, we would expect that the fraction of collisions with high relative kinetic energy in a reaction mixture will increase substantially with increasing temperature (even though the *average* relative molecular speed changes only slightly).

We are now in a position to write down the frequency of collisions leading to chemical reaction; it is as follows:

frequency of reactive collisions per unit volume = $Z_{AB} \exp(-E_0/RT)$ (110)

But this frequency is just a measure of the rate of reaction; that is,

$$-\frac{1}{V}\frac{dN_A}{dt} = Z_{AB} \exp(-E_0/RT) \quad (111)$$

where the left-hand side of the equation is the rate of change of the number of A molecules per unit volume. If we substitute for Z_{AB} from equations 106 and 107, and write the collision cross-section, σ, as πd^2, then equation 111 becomes

$$-\frac{1}{V}\frac{dN_A}{dt} = \pi d^2 \left(\frac{8kT}{\pi\mu}\right)^{1/2} \left(\frac{N_A N_B}{V^2}\right) \exp(-E_0/RT) \quad (112)$$

This equation looks complicated, but it can be written in a far more familiar form. The first (but perhaps not too obvious) step is to multiply both sides of the equation by the quantity $1/L^2$:

$$-\left(\frac{1}{L}\right)\left(\frac{1}{LV}\frac{dN_A}{dt}\right) = \pi d^2 \left(\frac{8kT}{\pi\mu}\right)^{1/2} \left(\frac{N_A}{LV}\right)\left(\frac{N_B}{LV}\right) \exp(-E_0/RT) \quad (113)$$

■ What is the quantity N_A/LV?

■ It is simply the concentration of A, [A], measured in mol m^{-3}. The number of moles of A, n_A, is given by N_A/L, and n_A/V is just the concentration of A; hence $N_A/LV = [A]$.

Since the quantities V and L are both constant, it is the case that

$$\frac{1}{LV}\frac{dN_A}{dt} = \frac{d}{dt}\left(\frac{N_A}{LV}\right) = \frac{d[A]}{dt} \tag{114}$$

Thus, equation 113 can be re-expressed as follows:

$$-\frac{d[A]}{dt} = L\pi d^2\left(\frac{8kT}{\pi\mu}\right)^{1/2}[A][B]\exp(-E_0/RT) \tag{115}$$

which may be written in the form:

$$-\frac{d[A]}{dt} = k_{\text{theory}}[A][B] \tag{116}$$

where k_{theory} is a theoretical rate constant (sometimes called the '**line-of-centres' rate constant**). The form of k_{theory} is

$$k_{\text{theory}} = L\pi d^2\left(\frac{8kT}{\pi\mu}\right)^{1/2}\exp(-E_0/RT) \tag{117}$$

Typical units for k_{theory} are m^3 mol^{-1} s^{-1}. To express k_{theory} in the more conventional units of dm^3 mol^{-1} s^{-1}, the value calculated from equation 117 must be multiplied by a factor of 10^3 (since 1 m = 10 dm).

■ What is the order of the theoretical rate equation given by equation 116?

▧ It is second order.

This, then, is the first conclusion we can draw from collision theory: for an elementary reaction involving two reactants, the order and molecularity are the same. You will recall that we simply stated this fact in our earlier discussion.

■ Do you recognize the form of equation 117?

▧ The exponential dependence on temperature of the theoretical rate constant is very similar to that found experimentally and expressed by the Arrhenius equation (equation 80), $k_R = A\exp(-E_a/RT)$.

This is our second conclusion from collision theory: we examine it in more detail below.

To summarize: the hard-sphere collision theory results in an equation which, qualitatively at least, has a firm basis in experiment and strongly supports the idea that a transformation involving an energy barrier to reaction is realistic. Next, we must examine the theory quantitatively.

9.1.3 The theory in practice

Collision theory provides no means of calculating the threshold energy, E_0: it is just a parameter, albeit a very important one, of the molecular model. However, if we assume that the threshold energy and the Arrhenius activation energy are of comparable magnitude, then by comparing equation 117 with the Arrhenius equation (equation 80) we can define a **theoretical A-factor**:

$$A_{\text{theory}} = L\pi d^2\left(\frac{8kT}{\pi\mu}\right)^{1/2} \tag{118}$$

It is this quantity that is compared with the corresponding experimental A-factor for an elementary reaction. (Strictly, there is no simple relationship between E_0 and the Arrhenius activation energy. The former is a parameter of a molecular model, whereas the latter is an experimental quantity and thus represents an average over many individual reactions. For present purposes, however, the assumption of equality between these two quantities is reasonable.)

You will notice that equation 118 predicts that the theoretical A-factor depends on the square root of temperature. In fact, for most reactions this dependence is completely swamped by the much stronger exponential temperature dependence. For example, you may recall from Section 6, that many reactions increase in rate by a factor of at least $2^{10} = 1\,024$, when taken from an ice-bath to a steam-bath; the predicted contribution of the A-factor to this increase is only a factor of $(373/273)^{1/2} \approx 1.17$. Experimentally, this would be extremely difficult to detect.

To calculate theoretical A-factors, we require values of collision diameters. We cannot obtain these directly, but can estimate them from experiments that measure properties that depend on molecular size. These are the properties that depend on the movement of molecules, for example viscosity of flow, heat conduction and diffusion of gases. Alternatively, they can be estimated from experiments on the way gases depart from ideal behaviour, or from diffraction measurements. Inevitably, these estimates are all different, and so there is some doubt about the size of the collision diameter, but not sufficient to cause concern about the simple theory.

Table 9 compares A-factors determined experimentally with those calculated from equation 118 for just a few bimolecular gas-phase reactions at 400 K.

   What conclusions can you draw from Table 9?

First, the calculated values of A-factors do not vary much from reaction to reaction, and are typically of the order of 10^{11} dm^3 mol^{-1} s^{-1}. The reason for this is easily seen. In equation 118 the only variables (at a fixed temperature) are the collision diameter, d, and the reduced mass, μ; the ratio $d^2/\mu^{1/2}$ does not change appreciably between one reaction and another. Second, the agreement between theory and experiment for the first reaction in the Table, while not good, is acceptable, considering the limitations of the theoretical model; but for the other reactions, particularly the last, the discrepancies are disturbingly large.

Why is the theoretical calculation so far out for some reactions? One way to highlight the difference is to modify equation 118 by introducing a variable parameter known as the **steric factor**, P:

$$A_{\text{theory}} = PL\pi d^2 \left(\frac{8kT}{\pi\mu}\right)^{1/2} \tag{119}$$

The inclusion of the factor P is justified on the grounds that even though a collision may be sufficiently energetic, reaction may not occur because the two molecules are not optimally orientated with respect to one another. This is a reasonable idea, but, unfortunately, collision theory provides no way of calculating P. It is thus given the empirical value, $P = A(\text{experiment})/A(\text{theory})$: some values are tabulated in Table 9. Defined in this way, the parameter clearly has no predictive value. Further, and more pertinent, it is hard to see how a term incorporating only geometrical factors can explain all of the large discrepancies encountered in practice.

Table 9 Comparison of experimental A-factors for bimolecular reactions with those calculated from collision theory at 400 K.

Reaction	$\dfrac{d}{\text{pm}}$	$\dfrac{A(\text{experiment})}{\text{dm}^3\,\text{mol}^{-1}\,\text{s}^{-1}}$	$\dfrac{A(\text{theory})}{\text{dm}^3\,\text{mol}^{-1}\,\text{s}^{-1}}$	P
$H_2 + Cl\cdot \longrightarrow HCl + H\cdot$	300	8.3×10^{10}	3.6×10^{11}	0.23
$NO_2 + O_3 \longrightarrow NO_3 + O_2$	390	6.3×10^9	1.7×10^{11}	0.04
$NO + O_3 \longrightarrow NO_2 + O_2$	350	8.0×10^8	1.6×10^{11}	0.005
$C_2H_4 + C_4H_6 \longrightarrow \textit{cyclo-}C_6H_{10}$	500	3.2×10^7	3.2×10^{11}	10^{-4}

To conclude, it is clear that the predictive powers of collision theory are very limited. None the less, as we have seen, it does provide a useful insight into the mechanism of bimolecular reactions.

SAQ 23 The bimolecular reaction:

$$CO + O_2 \longrightarrow CO_2 + O \tag{120}$$

was investigated in the gas phase in the temperature range 2 400 to 3 000 K and found to have an experimental A-factor of 3.5×10^9 dm^3 mol^{-1} s^{-1}. What is the steric factor for the reaction? (Assume the hard-sphere collision diameter is $d = 365$ pm, and take the average relative speed of CO and O$_2$ molecules at 2 700 K to be 1.957×10^3 m s^{-1}.)

9.1.4 An alternative approach: the reactive cross-section

Instead of focusing attention on 'hard spheres and steric factors', the collision theory can be used to introduce the concept of a **reactive cross-section**, σ_R. This cross-section represents the *effective target area* that results in a reactive collision between two species. In essence, the 'size' of the target area measures the propensity to react. Thus, we say molecule A has a small reactive cross-section in collision with molecule B when there is a small probability per collision of a reaction between A and B. Overall, the reactive cross-section is a complex quantity, because it incorporates all features, such as molecular forces, relative translational energy, orientation, etc., that together determine this area. It is thus an extremely difficult parameter to calculate from first principles. However, a great deal of information regarding its character can be determined experimentally; for example, from experiments involving **crossed molecular beams**.

A schematic diagram of a molecular beam apparatus is shown in Figure 23. Beams (the blue lines) of reactant molecules from two sources intersect at a 90° angle, and the products of reaction are detected both as a function of energy and of angular distribution. The beams are collimated, that is directed into fine parallel-sided beams, by passing through slits. Depending on the experiment, the beams can be of well-defined molecular speed, of selected internal energy state such as vibrational state, or even of selected molecular orientation. The whole experiment is carried out at very low pressures to ensure that collisions occur only in the reaction region: the experiment thus investigates *single collision processes*. An ideal molecular beam is so dilute that molecular collisions within the beam itself are precluded. The analysis and evaluation of molecular beam experiments are complex, and so a simplified description of just a few results will be given here.

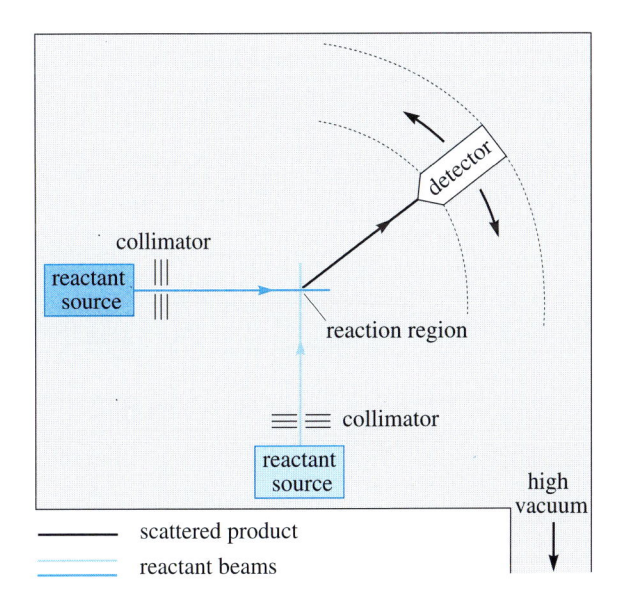

Figure 23 Schematic diagram of a crossed molecular beam apparatus, which is enclosed in a high-vacuum system.

When a beam of potassium atoms is crossed with a beam of iodine molecules, the following reaction occurs:

$$K + I_2 \longrightarrow I + KI \qquad (121)$$

The reactive cross-section was found to be of the order of 1.3 nm².

- What would you estimate the 'hard-sphere' collision cross-section to be? (Take d to be 300 pm.)

- Recalling that $\sigma = \pi d^2$, then
$\sigma = \pi \times (300 \times 10^{-12}\,m)^2 = 0.28 \times 10^{-18}\,m^2$
or, (since 1 m = 10^9 nm), $\sigma = 0.28$ nm²

What does the difference between the reactive and the 'hard-sphere' collision cross-section imply?

Because the experimental cross-section is much larger, it implies that the reaction between K and I_2 can occur at distances considerably greater than the hard-sphere collision diameter. The reaction is, in fact, an example of a **stripping reaction**: the K atom 'simply' strips off an I atom 'as it goes by', leaving the remaining I atom as a fairly disinterested spectator. A feasible explanation of this process is given by the so-called **harpoon mechanism**: at fairly large distances (600 to 700 pm) the alkali metal atom transfers an electron (the harpoon), which attaches itself to the I_2 molecule. Instead of two neutral species, there are now two charged species, K^+ and I_2^-, and they attract one another. This ionic attraction now provides the line for the alkali metal (the whaler) to haul in the halogen (the whale). Stable KI is then formed, and an I atom ejected. Reactions of alkali metal atoms (Li, Na, K, Rb, Cs) with halogens, and at least one of these metals with compounds such as BrCN, CBr_4, IBr, ICl, ICN and NOCl, have all been found to show characteristics of stripping reactions.

Molecular beam experiments can also be used to study steric effects directly. For example, for the reaction

$$K + CH_3I \longrightarrow CH_3 + KI \qquad (122)$$

it is possible to produce a beam of CH_3I molecules in which a fraction of the molecules are aligned in the same way. Analysis of the results shows that reaction is more probable when the CH_3I molecules in the beam are selected so that the potassium approaches the iodine end of the molecule rather than if alignment is such that the potassium approaches the methyl end. From these experiments a direct estimate of a steric factor can be made.

9.2 The solution phase: collisions and encounters

The molecular picture of a liquid is very different from that of a gas. Computer simulations, based on realistic interactions between molecules, provide a good picture of the arrangements of molecules in the liquid state. Such a simulation is shown in Figure 24, which is a cross-section through a liquid in which the motions of all the molecules have been instantaneously quenched; the picture is drawn on a molecular scale, and the molecules are assumed to be nearly spherical. The picture shows most of the molecules as being closely packed together – roughly as in a solid, but with an absence of long-range order. Hence, in a liquid, molecules must 'squeeze' past one another if they are to make substantial translational movements. By contrast, the movement of molecules in a gas is largely unhindered, and a comparable computer simulation (for one mole of gas at one atmosphere pressure) would show, at the most, only one or two molecules in the same cross-sectional area as in Figure 24.

It seems then that a reactant molecule in solution must 'weave' its way through a maze of molecules, most of which are molecules of the solvent, before it can reach a partner with which to react. This, in turn, suggests a physical picture in which a reactant molecule is surrounded by a cage of solvent molecules. This **solvent cage** is

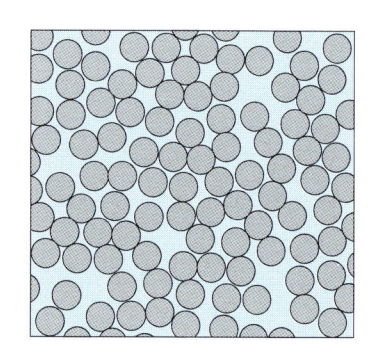

Figure 24 A computer simulation of an instantaneous configuration in a liquid.

a fairly 'loose' arrangement of molecules, but it is sufficient to trap a reactant molecule for a significant time. For example, for water at room temperature, the lifetime of a solvent cage is estimated to be of the order of 10^{-11} s (by comparison, a collision in the gas phase occurs on a time-scale of 10^{-13} s). While trapped in the solvent cage, the reactant molecule will undergo a sizeable number of collisions (estimates for water suggest about 100) with its nearest neighbours.

Eventually, the cage will open, due in part to the continual jostling motion of the molecules, and the reactant molecule escapes, but only to jump into another cage, and so on. The reactant thus migrates through the solution in a series of small, random, displacements: we refer to this as *diffusion*. Occasionally, however, a reactant molecule will slip into a cage containing another reactant molecule. When two reactants share the same cage they are known as an **encounter pair**. The events leading to the formation of an encounter pair between two molecules A and B are shown schematically in Figure 25. It is worth noting that, in general, the cage model accounts quite successfully for transport processes, such as diffusion, in pure liquids.

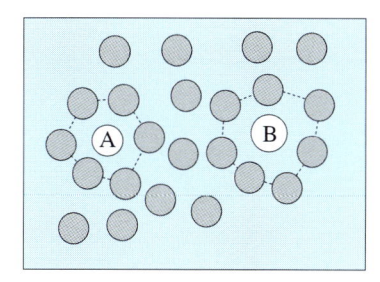

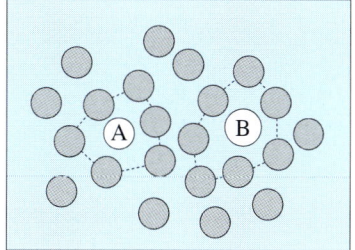

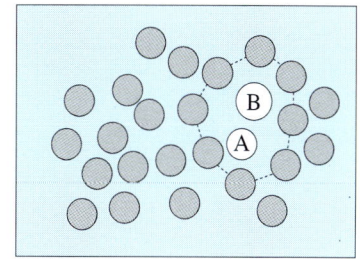

Figure 25 Schematic illustration of the events leading to the formation of an encounter pair between two molecules A and B in solution. (Solvent cages are indicated by broken lines.)

The picture outlined above suggests that an encounter between two reactant molecules in solution lasts considerably longer than a collision in the gas phase; equally, the cage effect results in a number of collisions occurring during each encounter. The pattern of collisions in the liquid phase, compared with the gas phase, is thus very different. In solution, occasional encounters occur, but with many collisions during one encounter, whereas in the gas phase the collisions are fairly evenly spaced in time.

It may seem then that the collision theory we have developed for the gas phase is inapplicable to solution reactions. However, consider a bimolecular reaction in solution with a relatively high activation energy, say 80 kJ mol^{-1}. Once an encounter pair is formed, its components, still identifiably reactant molecules, must overcome an energy barrier of roughly this magnitude if they are to form products.

■ What proportion of the total collisions in unit time will lead to reaction?

▪ The proportion will depend on the quantity $\exp(-E_a/RT)$; that is,

$$\exp\left(\frac{-80 \times 10^3 \text{ J mol}^{-1}}{8.314 \text{ J K}^{-1} \text{ mol}^{-1} \times 300 \text{ K}}\right) = 1.2 \times 10^{-14}$$

Thus, on average, it is only after 10^{14} collisions that a chemical transformation will occur. Clearly, the difference in pattern between collisions in the gas phase and collisions in solution will be unimportant in such a case.

For bimolecular reactions in solution with 'large' activation energies, we would therefore expect the *A*-factors to be similar to those predicted by the collision theory of gases: by 'large' in this context we mean activation energies greater than about 20 kJ mol^{-1}. The kinetics of such reactions are referred to as being **activation controlled**.

Table 10 collects together the Arrhenius parameters for a few bimolecular solution reactions. As predicted, the *A*-factors are quite close to the value expected from the collision theory of gases (that is, approximately 10^{11} dm^3 mol^{-1} s^{-1}). Notice also that for one of the reactions, the Arrhenius parameters are not significantly affected by a change of solvent. The results in the Table are, in fact, typical of a number of other solution reactions.

Table 10 Arrhenius parameters for some bimolecular reactions in solution.

Reactants	Solvent	$\dfrac{A}{dm^3\,mol^{-1}\,s^{-1}}$	$\dfrac{E_a}{kJ\,mol^{-1}}$
$CH_3Br + I^-$	water	1.66×10^{10}	76.3
$CH_3Br + I^-$	methanol	2.24×10^{10}	76.3
$CH_3Br + I^-$	ethane-1,2-diol	4.57×10^{10}	74.8
$CH_3I + S_2O_3^{2-}$	water	2.19×10^{12}	78.7
$CH_3I + C_2H_5O^-$	ethanol	2.42×10^{11}	81.6
$C_2H_5Br + OH^-$	ethanol	4.30×10^{11}	89.5

It is possible to develop a more quantitative description of an activation-controlled reaction. The approach, based on collision theory, pictures the encounter pair undergoing repeated collisions, while exchanging vibrational energy with the surrounding solvent molecules. In this way, the pair of molecules accumulates sufficient energy to overcome the energy barrier to reaction. For non-polar reactant molecules, the theory predicts an A-factor of the order of $10^{12}\,dm^3\,mol^{-1}\,s^{-1}$; that is, an order of magnitude larger than for a similar gas-phase reaction. Unfortunately, there are very few, if any, good examples of bimolecular reactions that occur both in the gas phase and in the solution phase against which this theory can be adequately tested.

> What will determine the reaction rate for a bimolecular reaction in solution that has a very low, or zero, threshold energy to reaction?

Reaction will occur at nearly every collision, and hence *at every encounter* in solution. Thus, it will be the frequency of encounters that sets an upper limit to the bimolecular reaction rate. As we have seen, this will depend on how rapidly the two reactants can diffuse through the solution. For this reason, such reactions are referred to as being **diffusion controlled**, and hence the rate of reaction at a given temperature is related to the viscosity of the solvent. It is possible to calculate the encounter rate using a model based on the laws that have been found experimentally to describe diffusion processes in bulk solutions. We shall not embark on the analysis here, but simply note that the model predicts that a second-order rate constant of between 10^9 and $10^{10}\,dm^3\,mol^{-1}\,s^{-1}$ is usually indicative of a diffusion-controlled reaction (but, as indicated above, this value also depends on the viscosity of the solution). If both reactants are charged, then the theory has to be modified to take account of interionic attractions or repulsions which, respectively, either enhance or restrict the diffusion of reactants together.

Diffusion-controlled reactions include recombination reactions of atoms and reactions between ions; a few examples are given in Table 11. These types of reaction are usually found to have a small activation energy. This activation energy reflects, not the energy barrier to chemical change, but rather the energy barrier associated with the diffusion of the reactant molecules through the solution. Diffusion is a thermally activated process, and for most solvents at room temperature, the activation energy for diffusion is typically of the order of $15\,kJ\,mol^{-1}$.

Table 11 Diffusion-controlled reactions in liquid solution at 298 K.

Reactants	$\dfrac{k_R}{dm^3\,mol^{-1}\,s^{-1}}$
$I\cdot + I\cdot$ (in hexane)	1.3×10^{10}
$H^+ + CH_3COO^-$	4.5×10^{10}
$OH^- + NH_4^+$	3.4×10^{10}

9.3 Summary of Section 9

1   Collision theory is based on a largely intuitive idea: chemical species must meet or collide before any chemical transformation can occur.

2   A more detailed treatment of gas-phase reactions is possible, because our knowledge of molecular properties in this phase is superior to that in solution.

In the gas phase

3   The rate of a bimolecular reaction depends on two factors: the collision frequency and the fraction of the total collisions that lead to chemical transformation:

- The *collision frequency*: molecules are assumed to be hard spheres; when spheres touch, a collision occurs. The collision frequency per unit volume, Z_{AB}, is then calculated starting from the model shown in Figure 21. To be realistic, the calculation takes into account the relative speeds of reactant molecules and the Maxwell distribution of this quantity.

- The *fraction of collisions*: two potentially reactive molecules will react only when the kinetic energy, as measured along the line joining their centres at impact, exceeds a critical threshold energy. The fraction of all collisions satisfying this criterion is given by $\exp(-E_0/RT)$.

4 The hard-sphere collision theory results in an expression for the rate of a bimolecular reaction that is similar to that found experimentally; in particular, the theoretical rate constant has an exponential dependence on temperature. Two important points are:

- The *predictive powers* of the theory are limited. The threshold energy cannot be calculated, and there are often large discrepancies between the theoretical and experimental values of the *A*-factor.
- An *empirical steric factor*, *P*, is introduced into the theory to allow for the orientational requirements of reaction.

5 Crossed molecular beam experiments can provide information on internal details of collision processes in gas-phase reactions.

In the solution phase

6 The cage model of a liquid can be used to describe the diffusion of reactant molecules towards one another: when two reactant molecules share the same cage they are known as an encounter pair.

7 The pattern of collisions in the solution phase is different from that in the gas phase. In solution, occasional encounters occur, with many collisions during each encounter.

8 For activation-controlled reactions (E_a greater than 20 kJ mol^{-1}), *A*-factors are similar in magnitude – typically 10^{10} to 10^{12} dm^3 mol^{-1} s^{-1} – to values predicted by the collision theory for gas-phase reactions.

9 For reactions with a small, or zero, threshold energy, reaction occurs at every encounter; these reactions are diffusion controlled and are expected to have second-order rate constants within the range from 10^9 to 10^{10} dm^3 mol^{-1} s^{-1}, the actual value being dependent on the viscosity of the solution.

10 TRANSITION STATE THEORY

As you have seen, collision theory provides a very direct physical picture of the mechanism of an elementary reaction: in order that molecules will react, they must collide with the correct orientation, and they must have a certain minimum energy. The theory is successful in so far as it accounts for the general form of the empirical Arrhenius equation, but its predictive powers are limited. It is possible, however, to develop a very different treatment of the events that occur when molecules collide during a bimolecular reaction. This is the **transition state theory**, or, as it is also known, the **activated complex theory** (the term 'absolute rate theory' is also occasionally used). The theory was first introduced by H. Pelzer and E. Wigner in 1932, and a few years later was considerably extended by Henry Eyring and his co-workers.

For gas-phase reactions, the theory treats molecules in a more realistic manner than simply representing them as hard spheres. This approach is precluded in solution, however, because of the complexities of the solvent interactions. Nevertheless, as you will learn in due course, the theory has the advantage that it can also be developed in a more general form, where thermodynamic concepts can be used: in this form it is particularly suited to a discussion of solution kinetics.

To introduce the basic ideas of transition state theory, it is first useful to examine the information that can be obtained from the construction of potential energy surfaces for simple gas-phase reactions.

STUDY COMMENT   You should plan to view video band 2 (*How do molecules react?*) during your study of Section 10.1. This sequence, in particular, should help you to visualize how potential energy surfaces are constructed in three dimensions.

10.1 Potential energy surfaces

10.1.1 An aside: potential energy curves

An isolated diatomic molecule can be pictured in simple terms as two nuclei, with a characteristic separation between them, surrounded by an electron cloud. Within the molecule there will be attractive forces between the electrons and the nuclei, and repulsive forces between the nuclei and also between pairs of electrons. These forces are balanced, so that the molecule has a well-defined stable structure: if the molecule is distorted in some relatively gentle way, it will return to this structure. Put in another way, we picture the molecule as being in a state of *stable equilibrium*. The manner in which the potential energy (or simply, stored energy) of the molecule varies as a function of internuclear distance, r, is described by a **molecular potential energy curve**: a schematic diagram of such a curve is given in Figure 26.

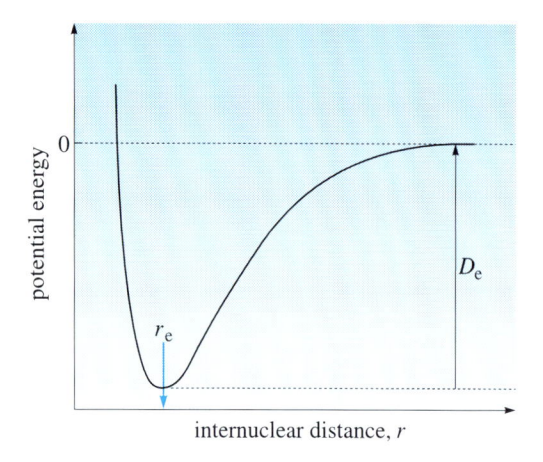

Figure 26 A schematic diagram of the molecular potential energy curve for an isolated diatomic molecule.

This curve, which has a characteristic shape, is often called a *Morse curve*, after the American physicist, Philip Morse, who fitted it using an empirical formula. The zero of energy is taken to be that of the two separated atoms at rest, in the limit that the internuclear distance tends to infinity: this choice will prove convenient in later discussions.

■ What is the significance of the minimum in the curve?

□ It corresponds to the most stable or *minimum energy* configuration of the two atoms. (In terms of our choice of the zero of energy, this occurs when the energy is at its most negative value.)

The internuclear distance for this most stable arrangement of the diatomic molecule is called the **equilibrium bond length**, and is given the symbol r_e.

■ Why does the potential energy rise sharply to the left of the energy minimum?

□ The sharp rise (the energy becomes less negative) is due to strong interatomic repulsive forces, which oppose the atoms moving closer together.

■ What happens to the right of the potential energy minimum?

□ To the right of the minimum, the attractive force between the atoms decreases as the internuclear distance increases, and hence the potential energy rises. At very large internuclear distances, the atoms behave as separate entities, because the attractive force between them is negligible.

The depth of the minimum, relative to the energy at infinite internuclear distance, is called the **dissociation energy** (or, sometimes, the **spectroscopic dissociation energy**). In this Course we shall give it the symbol D_e.

Using sophisticated quantum mechanical methods, the molecular potential energy for a diatomic molecule can be calculated. A few examples of these calculations, for selected diatomic molecules, are given in Table 12. The table compares computer-calculated (theoretical) values of the dissociation energy and the equilibrium bond length with the corresponding experimental values as determined by spectroscopic methods. It can be seen that the theoretical calculations are quite accurate for relatively simple diatomic molecules which have just a few electrons. Notice that the table lists values of the *molar* dissociation energy.

Table 12 Comparison between theoretical and experimental values of D_e and r_e for selected diatomic molecules.

Molecule	D_e/kJ mol^{-1} Theory	Experiment	r_e/pm Theory	Experiment
H_2	456	458	74.16	74.13
HF	600	589	92.0	91.7
F_2	161	162	141.0	142.0
O_2	455	503	122.0	120.7

Of course, we know from spectral evidence that diatomic molecules vibrate: the two atoms are not held rigidly apart at the distance of the equilibrium bond length, but vibrate back and forth about this separation. We also know that the molecule will rotate, but because rotational energy plays a comparatively small kinetic role we shall neglect it in this discussion.

> Can you recall from the Second Level Inorganic Course how the effect of vibration can be indicated on a potential energy curve such as that given in Figure 26?

The key point is that the energies of molecular vibrations are *quantized*. Figure 27 shows how we indicate this on the potential energy curve for the ground electronic state of the hydrogen molecule. The horizontal blue lines represent the allowed *total* energies of the vibrations; each **vibrational energy level** is labelled with a quantum number, denoted by the symbol v. The vibrational energy may change only by exactly the difference between two allowed energy levels. At the maximum extension or compression of the bond (in terms of a classical picture), represented by points at which the horizontal lines meet the Morse curve, the vibrational energy is wholly potential. In between, the energy is partly kinetic and partly potential.

■ What do you notice about the lowest vibrational energy level; that is, the $v = 0$ level?

▨ It does not coincide with the energy minimum. Thus, a diatomic molecule will always possess a minimum amount of vibrational energy, even at 0 K.

When the molecule has this minimum amount of energy it is referred to as being in its *vibrational ground state*: the vibrational energy in this state is called the **zero-point energy**. At normal temperatures and pressures, virtually all of the molecules in a diatomic gas will be in their vibrational ground state.

We have now introduced most of the concepts that we shall find useful in the discussion of potential energy surfaces. But there is one final, and very significant, piece of information that we can extract from Figure 27.

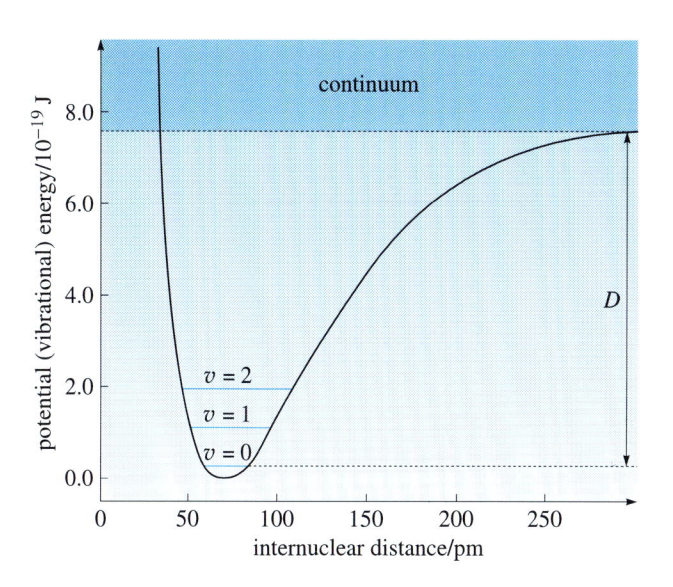

Figure 27 The ground-state potential energy curve for the hydrogen molecule, showing a few vibrational energy levels.

What is the energy barrier to decomposition of molecular hydrogen (initially in its vibrational ground state) to atomic hydrogen?

As a hydrogen molecule attains more vibrational energy (for instance, during a collision), the molecule will vibrate over greater internuclear distances; in simple terms, the vibration becomes more violent. If the molecule receives sufficient vibrational excitation, it will separate into atoms *within just one vibration*; that is, it will dissociate. The energy needed for this dissociation is the difference between the energy at which the vibrational levels reach a continuum of energy states, and the zero-point energy. This difference is represented by the symbol D in Figure 27 and is called the **thermochemical dissociation energy** or **bond dissociation energy**. (Comparison with Figure 26 shows that D is smaller than the spectroscopic dissociation energy, D_e, by an amount equal to the zero-point energy.)

The bond dissociation energy is a good measure of the energy barrier for the decomposition reaction. We thus see another important feature of Figure 27: *it provides a pictorial view of the energy barrier to an elementary reaction as a rise in a potential energy curve, which must be 'climbed' if a molecule is to dissociate.*

10.1.2 A simple example of an elementary reaction

The potential energy curve for the hydrogen molecule provides a pictorial view of the energy barrier that must be overcome if the molecule is to dissociate. We must now consider whether it is possible to extend this idea to a more interesting, but still relatively simple, *bimolecular* reaction. We shall choose the following gas-phase reaction as an example:

$$\text{F}\cdot + \text{H}-\text{H}' \longrightarrow \text{F}-\text{H} + \text{H}'\cdot \qquad (123)$$

We have labelled one of the hydrogen atoms with a prime; this is simply for convenience, so that it can be identified as the one that does *not* react with the fluorine atom.

What information about the reaction can be deduced from Table 12?

The table indicates that the *depth* of the potential well (measured with reference to a state in which the atoms are at infinite separation) is greater for a hydrogen fluoride molecule than for a hydrogen molecule. This fact must be reflected in any potential energy diagram we draw.

To construct a potential energy diagram, we consider equation 123 to represent a system consisting of three atoms: F·, H· and H'·. The potential energy of any arrangement of these atoms will depend on three internuclear distances, $r(\text{F---H})$, $r(\text{H---H}')$ and $r(\text{F---H}')$, where the distances between pairs of atoms are specified in brackets.

■ Can you see the difficulty in constructing such a potential energy diagram?

▨ To represent the energy of the system would require a plot in four dimensions – three distance coordinates and one energy coordinate. This is not practicable!

It simplifies matters to assume that all three atoms are arranged in a *linear* configuration *throughout the reaction*. We shall comment on this assumption later. This linear system has only two *independent* internuclear distances; the most convenient to choose are $r(\text{F---H})$ and $r(\text{H---H}')$. The potential energy of the system can then be represented as a surface in just three dimensions. An attempt at a schematic, perspective view of this surface, based on the ideas that we have developed so far, is shown in Figure 28.

In the Figure, the 'height' of the surface above a particular point, specified by the internuclear distances $r(\text{F---H})$ and $r(\text{H---H}')$, is the potential energy of the system (calculated relative to fully separated F· + H· + H'· atoms).

■ In the diagram, what does the face that is shaded darker blue represent?

▨ It is a section through the surface perpendicular to the $r(\text{F---H})$ axis. This face therefore represents the *potential energy curve* for the molecule H—H' as the fluorine atom approaches.

Similarly, the light blue face represents the potential energy curve for the molecule F—H as the other hydrogen atom (H') moves away.

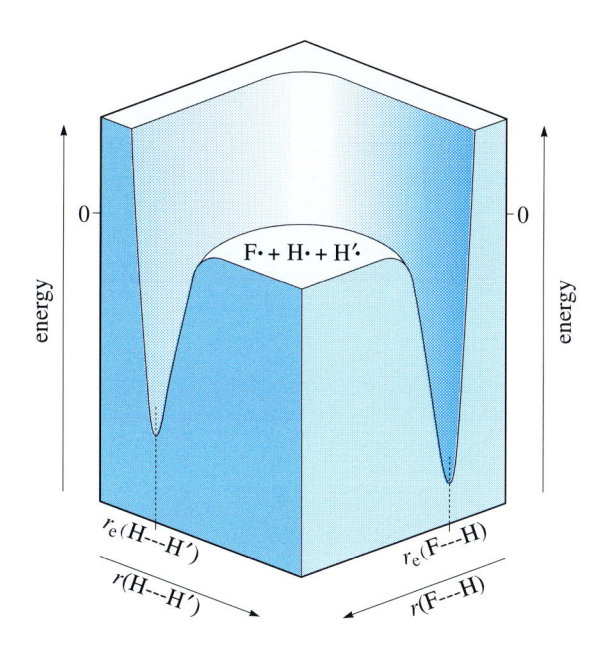

Figure 28 A schematic view of the potential energy surface for the linear system F---H---H´.

Clearly, chemical transformation must take place in the region of the potential energy surface where r(F---H) and r(H---H´) are both of the order of chemical bond lengths. How do we obtain information about this region?

A very detailed quantum mechanical calculation has been carried out, from first principles, for around 150 different linear configurations of the fluorine atom plus hydrogen molecule system. The **potential energy surface** derived from these calculations is believed to be quite realistic.* So, to answer our question, let us look at this surface.

Figure 29 shows two representations of computer-drawn perspective views of the potential energy surface. An animated view of how the surface is constructed in three dimensions is provided in video band 2.

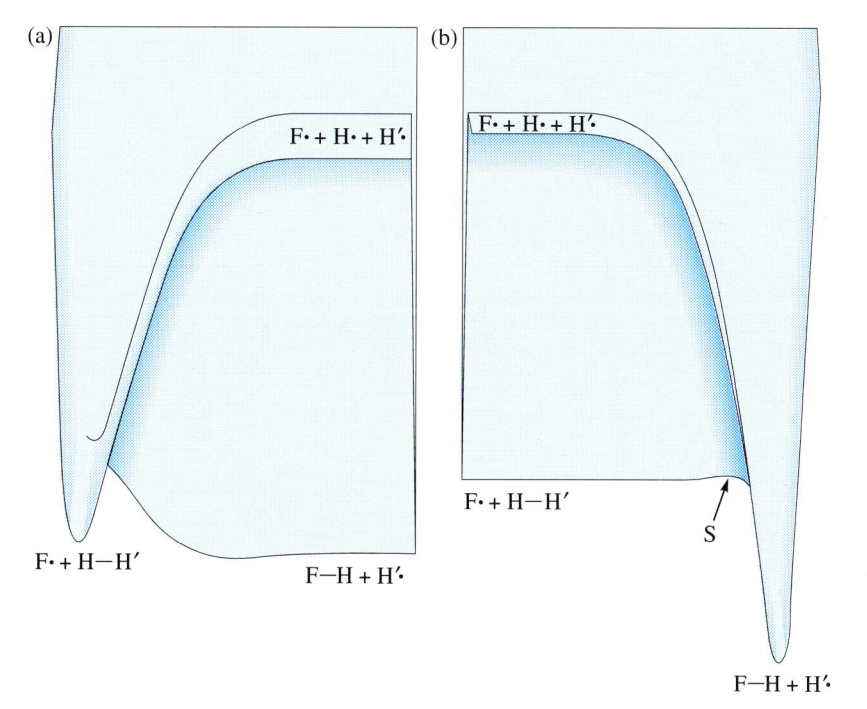

Figure 29 Representations of two computer-drawn perspective views of the potential energy surface for the linear system F---H---H´. (The significance of the label S in part (b) will become apparent shortly.) You may find it helpful to view these diagrams from a distance.

* The reaction F• + H$_2$ ⟶ HF + H•, apart from being of intrinsic interest, forms the basis for the HF chemical laser, since the HF produced in the reaction has an inverted population of vibrational energy levels which makes a lasing action possible. Attempts to define and determine the potential energy surface have a long history, beginning in 1931. All available theoretical and computational methods have been used since that time. The particular surface we examine in this Section is based on a least-squares fit to the *ab initio* 'BOPS' surface, so-called after C. F. Bender, S. V. O'Neill, P. K. Pearson and H. F Schaeffer III, who reported it in *Science*, volume 176, pp. 1412–1414 (1972).

It is customary to describe a potential energy surface in 'geographical' terms, the whole surface being viewed as a sort of 'chemical landscape'. Hence, Figure 29a shows the **entrance valley** for the reactants, F· + H—H´, and Figure 29b shows the **exit valley** for the products, F—H + H´·. However, it is difficult to extract quantitative information from such representations; it is, in fact, more useful to plot a **contour map**, as in Figure 30.

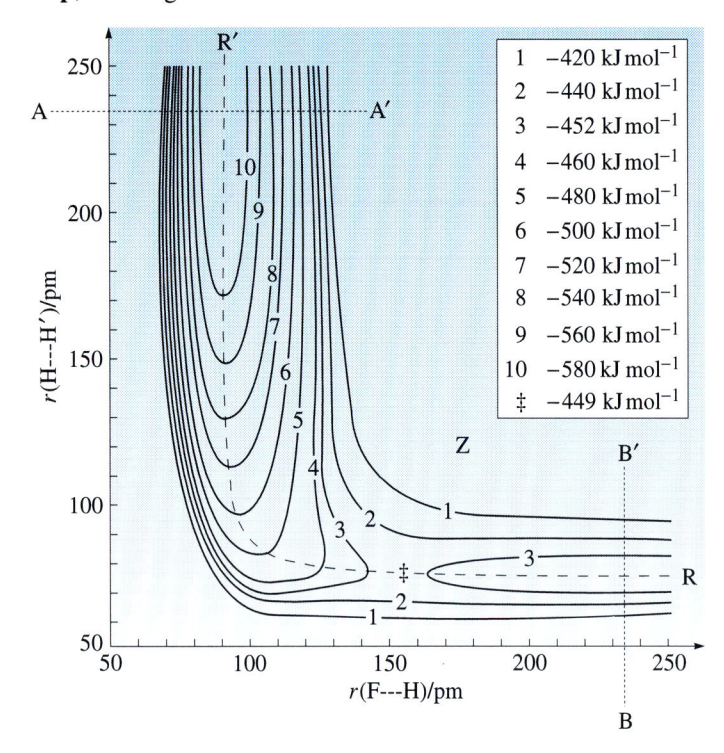

Figure 30 A contour map of the linear F---H---H´ potential energy surface. The zero of energy is taken to be that of completely separated F·, H· and H´· atoms. Contours are drawn at energies as labelled on the Figure.

Each curved line (energy contour) in Figure 30 represents all the points – each characterized by a particular pair of internuclear distances r(F---H) and r(H---H´) – at which the potential energy has the particular value (measured with respect to a state in which all three atoms are at infinite separation) indicated in the Figure. To continue the geographical analogy, the contour map is equivalent to an Ordnance Survey map of the chemical landscape depicted in Figure 29. For convenience, the contours are labelled in kJ mol^{-1} since molar energies are easier to handle: *strictly, the diagram represents the interaction of* **one** *hydrogen molecule with a* **single** *fluorine atom.*

SAQ 24 This question refers to the contour map given in Figure 30. Describe the arrangement of atoms, F·, H· and H´· at the position labelled Z, and also indicate the arrangement of atoms that corresponds to cross-sections taken along the dotted lines A·······A´ and B·······B´.

> Your answer to SAQ 24 has described the limiting regions of the contour map. What happens in the region where chemical transformation takes place?

The three atoms F·, H· and H´· are close together (as in a collision of H—H´ with F· or H´· with H—F) in the region of the potential energy surface where the entrance and exit valleys intersect. In this region there is a **saddle point** (or **col**), so called because the curvature of the surface corresponds to that of a saddle (or mountain col). In Figure 30, the saddle point is marked by the symbol ‡ (spoken as 'double dagger').

> What is the significance of the broken line R---R´ in Figure 30?

It represents the **minimum-energy path** that the reaction can take over the potential energy surface. To continue the geographical analogy, it represents a pass between two valleys. Along this path the saddle point corresponds to a maximum in energy. This maximum can just be seen in Figure 29b, where it is marked by the letter S. (It is difficult to discern because, as you will find shortly, the activation energy for the forward reaction is relatively small.) Notice that, conversely, the saddle point corresponds to a minimum in energy for a path passing through it, but at right-angles to the minimum-energy path.

Any path that the system may take in passing from reactants to products is called a **reaction coordinate**. A two-dimensional cross-section through the potential energy surface along a reaction coordinate shows how the energy changes along this particular path: it is referred to as a **potential energy profile**. The energy profile along the minimum-energy path is of special importance: it is shown in Figure 31.

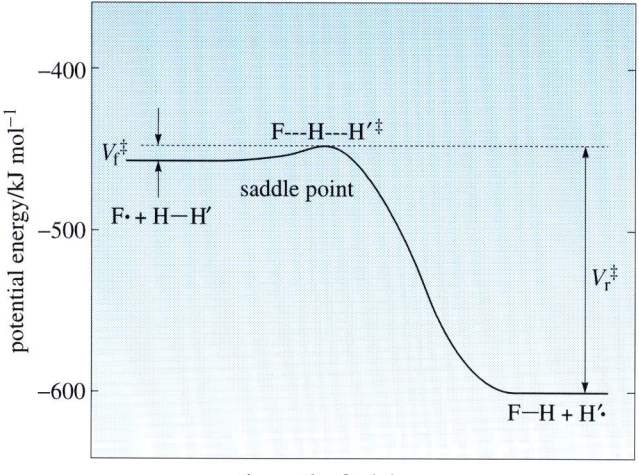

Figure 31 The potential energy profile along the minimum-energy path for the reaction F· + H—H′ ⟶ F—H + H′·. The subscripts f and r refer to the forward and reverse reactions, respectively.

The energy profile shows quite clearly that there is an energy barrier to the forward reaction, albeit a small one. The barrier height, which is labelled $V_f^\ddagger$ in Figure 31, is the energy difference between the saddle point and the initial state of the system. It has a calculated value of 7.0 kJ mol^{-1}.

- According to Figure 31, what is the energy barrier for the reverse reaction?

$$F-H + H'\cdot \longrightarrow F\cdot + H-H' \tag{124}$$

- It is the energy difference between the saddle point and the final state of the system, which is labelled $V_r^\ddagger$ in the Figure. The calculated value is 151 kJ mol^{-1}.

The state of the system at the top of the energy barrier is called the **transition state**. In the transition state, all three atoms are regarded as being partially bonded in a high-energy configuration, which we represent as F---H---H′‡. This species is called an **activated complex**.

Do you think that the activated complex can be regarded as a stable species?

The activated complex is at the maximum of energy along the reaction coordinate shown in Figure 31, but it is also at the minimum of energy for motion perpendicular to this coordinate. Thus, in a sense, it is both unstable and stable; it is best regarded as having a definite, but entirely transitory, molecular configuration. It must be emphasized that an activated complex is *not* to be confused with a reaction intermediate – these, as will be made clear in Block 3, are either products or reactants in elementary steps in a reaction mechanism, and have a lifetime much longer than any conceived for the activated complex.

We should stress that Figure 31 corresponds to a linear configuration of all three atoms throughout the reaction. The calculation can be extended to other configurations; for instance, for a perpendicular approach of F· to H—H′, the energy barrier is found to increase by about 50 kJ mol^{-1}. However, the colinear configuration has the smallest calculated potential energy barrier to reaction, and so we would expect it to be the one most favoured in practice; a conclusion that is borne out by experiment, as we shall see later.

It is tempting to equate the energy barrier, $V_f^{\ddagger}$, in Figure 31 with the experimental activation energy. But is this correct? Before answering the question, we should recall that all molecules, including an activated complex, must possess at least zero-point vibrational energy. Hence, strictly, before making any comparison with experiment, the barrier height should be corrected for the difference in zero-point energy between the activated complex and the hydrogen molecule. This corrected barrier height is called the **zero-point internal energy of activation**, and is given the symbol $\Delta E_0^{\ddagger}$. Since $\Delta E_0^{\ddagger}$ is the minimum energy that must be supplied to the reactants to achieve the transition state, it is equivalent to the threshold energy that we introduced in collision theory in Section 9.1.2. The calculated value of $V_f^{\ddagger}$ is 7.0 kJ mol^{-1}, and the correction for zero-point energy can be estimated to be +2.8 kJ mol^{-1}, so that $\Delta E_0^{\ddagger}$ = 9.8 kJ mol^{-1}. A value of the experimental activation energy (which is very difficult to measure) is 7.1 kJ mol^{-1}. Thus, to answer the question (despite the fact that $\Delta E_0^{\ddagger}$ is derived from a molecular model, whereas the activation energy is an experimental quantity reflecting a complex average over a multitude of individual reactions), it appears reasonable to accept that the theoretical and experimental values are of comparable magnitude. Of course, for this conclusion to be valid, the 'reality' of the potential energy surface must be critically assessed. Without going into detail, it is accepted that the surface probably approaches 'chemical accuracy'; that is, it is valid to within about 4 kJ mol^{-1}.

> What is the significance of the energy difference between reactants and products in Figure 31?

It is the difference between the energy required to break the H—H' bond and that *released* when the H—F bond is formed; that is, D_e(H—H') – D_e(H—F). If we took the effect of zero-point vibrational energy into account, then the energy difference would be just the *internal energy change* occurring for the reaction at 0 K. However, it is a reasonable approximation to equate this internal energy change with an enthalpy change, which may be calculated from thermodynamic data at 298.15 K.

■ Use information from the S342 *Data Book* to calculate $\Delta H_m^{\ominus}$ at 298.15 K for reaction 125, that is (without labelling the individual hydrogen atoms)

$$F\cdot + H_2 \longrightarrow HF + H\cdot \qquad (125)$$

■ $\Delta H_m^{\ominus} = \Delta H_f^{\ominus}(HF,g) + \Delta H_f^{\ominus}(H,g) - \Delta H_f^{\ominus}(F,g) - \Delta H_f^{\ominus}(H_2,g)$

Using values from the S342 *Data Book*:

$\Delta H_m^{\ominus} = (-271.1 + 218.0 - 79.0 - 0) = -132.1 \text{ kJ mol}^{-1}$

This value compares favourably with that (–144 kJ mol^{-1}) determined in the potential energy surface calculation, which again gives us confidence in the realistic nature of the calculation.

There is one final but important point. It should be recognized that the *whole* of a calculated potential energy surface is of interest in examining the *dynamics* of a collision between potentially reactive chemical species. The motion, or *trajectories* of reactants or products, and the important role of vibrational energy in elementary reactions, can be visualized using these surfaces. As a consequence, a much deeper knowledge can be gained of the processes that occur at the very heart of chemical reactions.

To summarize: our examination of the calculated potential energy surface of a 'simple' bimolecular reaction has provided a great deal of insight into the energy changes that occur when reactants are converted into products during an elementary reaction. In particular, the calculated energy barrier to reaction is very similar in magnitude to the experimentally determined activation energy.

10.1.3 Other elementary reactions

For many bimolecular reactions, particularly those in solution, it is not possible to calculate a potential energy surface from first principles, and hence derive an energy profile for the reaction. Nevertheless, the idea of a minimum energy path for reaction is a very useful concept, and it is customary practice to draw *schematic* energy profiles for elementary reactions. Such a profile is drawn in Figure 32 for an exothermic reaction. These profiles are taken to have the same characteristics as those that would be determined by detailed calculation.

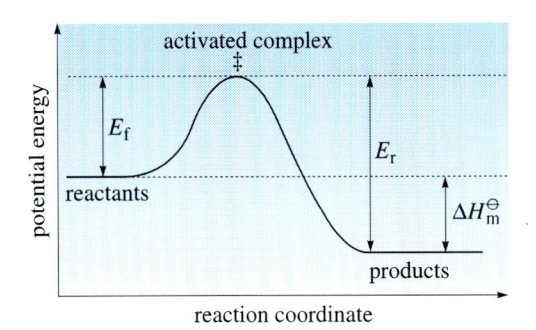

Figure 32 A schematic energy profile for an exothermic elementary reaction. (The quantities E_f, E_r and $\Delta H_m^\ominus$ are defined in equation 126.)

■ In Figure 32, the quantities E_f and E_r represent the experimental activation energies for the forward and reverse reactions, respectively: they are both positive quantities. What is the relationship between these quantities and the enthalpy change for the overall reaction (which is negative)?

□ It should be clear from the diagram that:

$$E_f - E_r = \Delta H_m^\ominus \tag{126}$$

Notice that the single fact that an elementary reaction is exothermic can tell us nothing about the energy barrier to the forward reaction (but if E_r is known then E_f can be estimated, or vice versa).

SAQ 25 Sketch an energy profile for an endothermic elementary reaction. What does this profile tell you about the energy barrier to reaction?

There is one final point about the schematic energy profile in Figure 32: it shows quite unequivocally that an elementary reaction must pass through only a *single* transition state. This leads to a very common definition of such a reaction:

> An elementary reaction is one that occurs in a single step and passes through a single transition state.

10.2 Transition state theory in the gas phase

Transition state theory provides a framework for predicting the rates of elementary reactions in which the concepts of a minimum-energy path to reaction, and an activated complex, are of central importance. To see how to develop the theory, we can consider a bimolecular gas-phase reaction between an atom and a diatomic molecule:

$$A + BC \longrightarrow AB + C \tag{127}$$

■ What is the rate equation for this reaction?

□ The reaction is elementary, so that

$$J = k_R[A][BC] \tag{128}$$

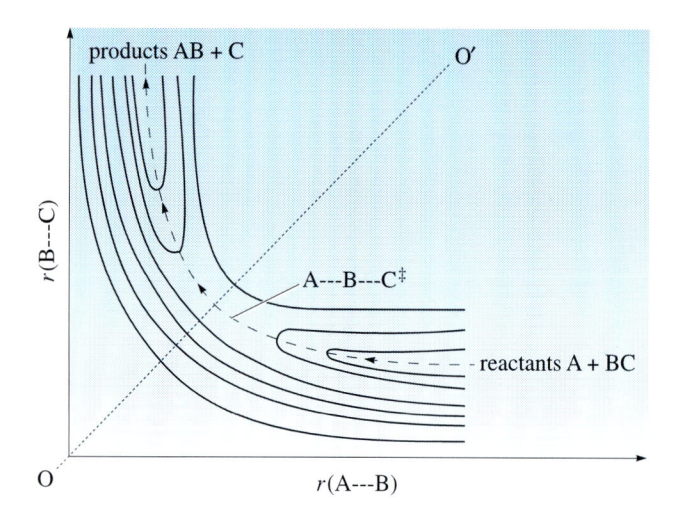

Figure 33 A schematic contour map of the potential energy surface for the reaction A + BC ⟶ AB + C.

The mechanism of the reaction involves, as we have seen, the formation of an activated complex, A---B---C‡. If the reaction is exothermic, the energy profile will be similar to that in Figure 32, and the activated complex will have a molecular configuration corresponding to the top of the energy barrier. A contour map of a hypothetical potential energy surface for the reaction is sketched in Figure 33, for a colinear approach of the reactants.

We might expect that the concentration of the activated complex will be very small compared with the concentrations of reactants, so that the overall rate of reaction will depend on the rate at which activated complexes can 'fall apart' to yield products. An alternative way of writing the rate equation is then:

$$J = v^{\ddagger}[\text{A---B---C}^{\ddagger}] \tag{129}$$

where [A---B---C‡] is the concentration of the activated complex, and $v^{\ddagger}$ gives the **frequency of decomposition** of the activated complex to give products. (We shall return to justify the form of this equation in more detail in due course.) Two problems now arise: first, it is not possible to measure [A---B---C‡] experimentally, and second, we must relate $v^{\ddagger}$ to some property of the activated complex.

To begin with, let us look at how to establish the concentration of the activated complex. It seems reasonable that this should be related to the concentrations of the reactants since, after all, it is the reactants that 'come together' to form the activated complex. For the moment, let us consider that the process

$$\text{A} + \text{BC} = \text{A---B---C}^{\ddagger} \tag{130}$$

maintains equilibrium *while the reaction proceeds*: in other words, we assume that the rate at which activated complexes 'leak away' to form products is not sufficient to disturb this equilibrium.

> Given this assumption, what is the relationship between [A---B---C‡] and the concentrations of reactants?

The concentrations of the reactants and the activated complex are related by an *equilibrium constant*, $K_c^{\ddagger}$

$$K_c^{\ddagger} = \frac{[\text{A---B---C}^{\ddagger}]}{[\text{A}][\text{BC}]} \tag{131}$$

which has dimensions of (concentration)$^{-1}$.

Combining equations 129 and 131 gives

$$J = v^{\ddagger} K_c^{\ddagger} [\text{A}][\text{BC}] \tag{132}$$

which, by comparison with equation 128, provides the following expression for the *theoretical* second-order rate constant:

$$k_{\text{theory}} = v^{\ddagger} K_c^{\ddagger} \tag{133}$$

You may be puzzled: in effect, all we have done is replace one uncertain parameter, [A---B---C‡], by another, $K_c^{\ddagger}$; the reason for this will become clear shortly.

However, before we go further, the equilibrium assumption embodied in the derivation of equation 133 requires some comment. It implies that the activated complex, *even though of a transitory nature*, can be regarded as a true thermodynamic entity, in equilibrium with reactants at the temperature of the reaction. Such a proposal has provoked widespread discussion. Suffice it to say that a far more sophisticated approach to the derivation can be taken that does not rely on this assumption. It shows that the assumption is satisfactory as long as the reacting system does not lurch so rapidly from reactants to products that even thermal equilibrium among reactant molecules cannot be maintained. The simplest, though not the most satisfying, justification of the overall approach is that calculations based on it are quite successful.

The value of the equilibrium constant $K_c^{\ddagger}$ in equation 133 can be expressed in terms of the *molecular properties* of both the reactants and the activated complex, using the methods of statistical mechanics. Unfortunately, these methods are quite complex and would take considerable space to develop here. However, our aim is to outline transition state theory, rather than analyse it in detail, and so we shall continue by simply highlighting the main points in the development of the theory. This means that we shall have to ask you to accept, without proof, one or two fundamental expressions. (You will not be expected to remember these in detail.)

Statistical mechanics evaluates $K_c^{\ddagger}$ as follows:

$$K_c^{\ddagger} = L \frac{Q^{\ddagger}}{Q_{\text{reactants}}} \exp\left(-\frac{\Delta E_0^{\ddagger}}{RT}\right) \tag{134}$$

The quantity $\Delta E_0^{\ddagger}$ is just the zero-point internal energy of activation, which we defined in the last Section: remember that it is a measure of the height of the energy barrier between activated complex and reactants. The quantities $Q^{\ddagger}$ and $Q_{\text{reactants}}$ are parameters that *in principle* can be calculated from a knowledge of the molecular detail and properties of the activated complex and the reactant molecules, respectively. Strictly, $Q^{\ddagger}$ is the **molecular partition function per unit volume** for the activated complex, and $Q_{\text{reactants}}$ is equal to the *product* of the molecular partition function per unit volume for each of the reactant molecules.

At a given temperature, a molecular partition function describes how the energy available to a molecule is distributed among the various translational, vibrational, rotational, and in certain cases electronic, energy levels of the molecule. For our purposes, the important point is that the value of a molecular partition function can be calculated from standard formulae (to be found in most advanced physical chemistry textbooks), which require only a knowledge of the geometry, mass, vibrational parameters and rotational parameters of the molecule at a stated temperature.

According to a statistical mechanical treatment, the transition state theory expression for the theoretical rate constant now becomes:

$$k_{\text{theory}} = v^{\ddagger} L \frac{Q^{\ddagger}}{Q_{\text{reactants}}} \exp\left(-\frac{\Delta E_0^{\ddagger}}{RT}\right) \tag{135}$$

As a final step, we need to relate the quantity $v^{\ddagger}$ to some property of the activated complex. How do we achieve this?

The criterion for the activated complex to pass over the energy barrier to give products is that there must be some form of motion along the reaction coordinate. You will recall (Section 10.1.1) that for the decomposition of a diatomic molecule, we identified this motion as being vibrational in character. Can we use a similar model for the activated complex?

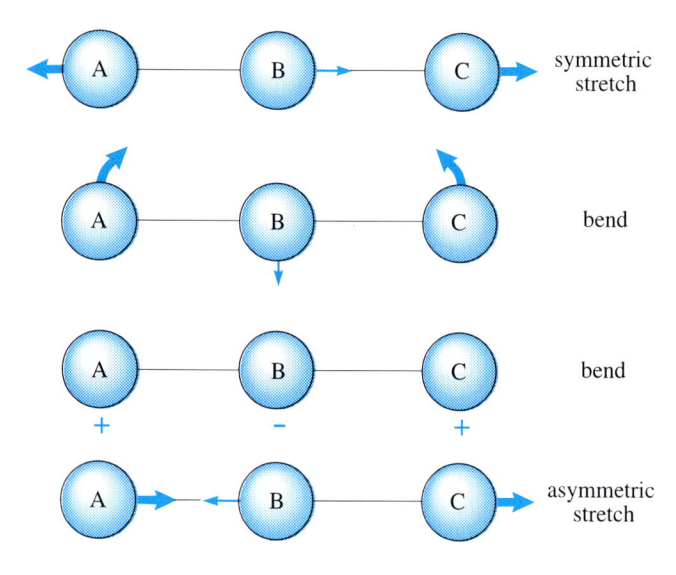

Figure 34 Vibrational modes of a stable, linear, triatomic molecule ABC. The labels '+' and '−' correspond to movements of the atoms normal to the plane of the page.

A *stable*, linear, triatomic molecule has four vibrational modes associated with it: two stretching and two bending. These are illustrated in Figure 34. We must now ask whether a linear, triatomic *activated complex* can also possess these modes.

In the symmetric stretch, both r(A---B) and r(B---C) extend and contract at the same time – hence the term symmetric. If you look at the contour map in Figure 33, you will see that this vibrational mode corresponds to motion along the line O-------O'. But O-------O' is perpendicular to the minimum-energy path reaction coordinate, and a cross-section along this line through the potential energy surface would reveal a potential energy *minimum* at the saddle point. Thus, any distortion of the activated complex along O-------O' will be subject to a 'restoring force', and hence the symmetric stretch vibrational mode is a 'true' vibrational mode.

■ What can we say about the bending vibrational modes?

□ We cannot discuss these in terms of the contour map in Figure 33, because this map corresponds to a *colinear arrangement* of atoms A, B and C throughout the reaction. However, it does turn out that bending modes of the activated complex can also be considered as true vibrational modes.

Hence, we are left with just the asymmetric stretch to consider. In this mode, r(A---B) contracts as r(B---C) extends, and vice versa. This is equivalent to motion along the minimum-energy reaction coordinate. But along this coordinate the activated complex is situated at the top of an energy barrier, so that there are no restoring forces for an asymmetric stretching vibration. *This vibration is thus not a true vibration because it leads to dissociation of the activated complex.*

Our simple discussion of the vibrational modes of the activated complex leads to the idea that there is a crucial vibration that corresponds to passage of the activated complex over the energy barrier to give products.

The frequency of this crucial vibration, which can be labelled v_c, will be equal to the frequency of the decomposition of the activated complex; that is, $v^{\ddagger}$, as defined in equation 129. In reality, there is a possibility that not every oscillation will take the activated complex through the transition state to give products. However, for our purposes, it is sufficient to assume that $v^{\ddagger} = v_c$.

The activated complex is very weakly bound with respect to decomposition, and so the frequency, v_c, will be much lower than that expected for an 'ordinary' molecular vibration. For this reason, it is possible to separate out from the overall partition function for the activated complex, $Q^{\ddagger}$, the particular contribution made by the vibration corresponding to v_c. In fact, it turns out that

$$Q^{\ddagger} = \frac{kT}{hv_c} Q^{\ddagger'} \qquad (136)$$

where h is the Planck constant, k is the Boltzmann constant, and the prime on the quantity $Q^{\ddagger'}$ signifies that one vibrational term is missing.

We are now in a position to pull the various strands of the calculation together, and to obtain a final expression for the theoretical rate constant. Combining equations 135 and 136 gives

$$k_{\text{theory}} = v^{\ddagger} L \frac{kT}{h v_c} \frac{Q^{\ddagger'}}{Q_{\text{reactants}}} \exp\left(-\frac{\Delta E_0^{\ddagger}}{RT}\right) \qquad (137)$$

or, assuming that $v^{\ddagger} = v_c$ as discussed earlier, and writing in a more compact form:

$$k_{\text{theory}} = \frac{kT}{h} K' \qquad (138)$$

where $K' = L(Q^{\ddagger'}/Q_{\text{reactants}}) \exp(-\Delta E_0^{\ddagger}/RT)$. Equation 138 is often called the **Eyring equation**. (The quantity $kT/h = 6.25 \times 10^{12} \text{ s}^{-1}$ at 300 K.)

If we move to reactions involving more than three atoms, the simple, rather pictorial, description of the theory outlined above is no longer feasible; for instance, reaction takes place on a multi-dimensional potential energy surface. However, the *general* form of equation 138 is still taken to be valid.

How can we compare transition state theory with experiment? Notice that the form of equation 137 is similar to that of the Arrhenius equation. It indicates that the theoretical rate constant is equal to an exponential term multiplied by a pre-exponential factor, which is almost independent of temperature when compared with the dominant temperature dependence of the exponential term. (Both $Q^{\ddagger'}$ and $Q_{\text{reactants}}$ are temperature dependent, but in such a way that cancellation of similar factors leaves only a slight temperature variation of their ratio.)

We have already indicated that $\Delta E_0^{\ddagger}$ may, to a good approximation, be equated with the experimental activation energy, so we can identify the pre-exponential term with a theoretical A-factor:

$$A_{\text{theory}} = \frac{kT}{h} L \frac{Q^{\ddagger'}}{Q_{\text{reactants}}} \qquad (139)$$

As in collision theory, it is this theoretical A-factor that we compare with experiment. But can the quantities in the equation be evaluated?

Of course, the fundamental constants and the temperature of the reaction will be known. Also, providing that the geometries and vibrational frequencies of the reactant molecules are available, the value of $Q_{\text{reactants}}$ can be determined from standard formulae. Spectroscopic studies can provide the information about molecular geometries and vibrations, although for larger molecules the interpretation of such spectra is not easy. The calculation of $Q^{\ddagger'}$ is even more problematic.

In most cases, the activated complex cannot be detected experimentally. Very modern laser experiments, as mentioned in Section 7.2, have been developed that can probe chemical processes on the femtosecond timescale. This is leading to the development of a whole new area of chemistry called **femtochemistry**. Femtochemical investigations have, in special cases, provided direct information on activated complexes, and they may find a wide range of applications in the future.

If a potential energy surface can be calculated, and it is one in which there can be some confidence, then information on the structure of the activated complex and its properties can be deduced. But there seems to be a danger here of the argument becoming circular. An accurate knowledge of the potential energy surface for a reaction is crucial to the detailed application of transition state theory – it is required to determine $\Delta E_0^{\ddagger}$, to identify the critical vibration along the reaction coordinate, and to estimate the properties needed to calculate $Q^{\ddagger'}$. And yet, as we have seen for most reactions, a potential energy surface cannot be calculated. Is transition state theory therefore simply an elegant theory that provides no quantitative predictions?

A compromise is required. The properties of the activated complex can be *estimated*, for instance by making comparison with similar *stable* molecules. Even then, the choice of reaction coordinate is still very much a matter of guesswork. However, this approach can be quite successful, as Table 13 shows.

Table 13 Comparison of *A*-factors calculated using transition state theory with experimental values.

Reaction	$A/\text{dm}^3\ \text{mol}^{-1}\ \text{s}^{-1}$ Experiment	Theory
Br· + H$_2$ ⟶ HBr + H·	3×10^{10}	1×10^{11}
H· + CH$_4$ ⟶ H$_2$ + ·CH$_3$	1×10^{10}	2×10^{10}
H· + C$_2$H$_6$ ⟶ H$_2$ + ·C$_2$H$_5$	3×10^9	1×10^{10}
·CH$_3$ + H$_2$ ⟶ CH$_4$ + H·	2×10^9	1×10^9
·CH$_3$ + CH$_3$COCH$_3$ ⟶ CH$_4$ + ·CH$_3$COCH$_2$	4×10^8	1×10^8
2ClO· ⟶ Cl$_2$ + O$_2$	6×10^7	1×10^8
F$_2$ + ·ClO$_2$ ⟶ FClO$_2$ + F·	3×10^7	8×10^7

The agreement between theory and experiment in the Table is considerably better than that achieved by collision theory (see Table 9, Section 9.1.3). In particular, the theoretical values of the *A*-factor vary quite considerably according to the complexity of the reactant molecules: there is no need to introduce arbitrary steric factors to explain large discrepancies between theory and experiment. It is the partition function quotient in equation 139 that allows the inclusion of 'molecular complexity' in the calculation, as illustrated in the following SAQ.

SAQ 26 It is possible to make an order of magnitude calculation of the ratio $Q^{\ddagger}/Q_{\text{reactants}}$ appearing in equation 139 for an elementary gas-phase reaction between two complex molecules (not necessarily containing the same number of atoms); at 400 K the value is $10^{-33}\ \text{m}^3$. What is the value of the theoretical *A*-factor at this temperature? How does this prediction compare with that for collision theory? Take as an example the last reaction in Table 9.

It is well to pause at this stage and consider where our examination of transition state theory has led us. On its introduction in the 1930s, the theory was greeted with enthusiasm, because it was confidently expected to be far superior to collision theory in predicting the rates of elementary reactions. However, as we have seen, the predictive powers of the theory, given that its inherent assumptions are accepted, depend critically on a knowledge of the properties of the activated complex. This is a major drawback of the theory, because there are only a handful of reactions for which the properties can be calculated in any detail; for other reactions, estimates of one form or another have to be made.

The importance of the theory to the practical kineticist lies in its framework; the concepts of an energy profile, a minimum-energy reaction path and an activated complex are frequently used in discussion of reaction rate. Indeed, in this Course we shall adopt these ideas on a variety of occasions. It is particularly useful in such circumstances to reformulate the theory in more general thermodynamic terms.

10.3 Thermodynamic aspects

An attractive feature of transition state theory is that it may be developed in a form that incorporates thermodynamic parameters. This is particularly important when examining solution reactions, because a molecular description of an activated complex surrounded by solvent molecules would be prohibitively complex. It is in a 'thermodynamic form' that the theory is most often used in practice.

The departure point for the more general development – for either bimolecular gas-phase or solution reactions – is the expression for the theoretical rate constant given by equation 133:

$$k_{\text{theory}} = v^{\ddagger} K_c^{\ddagger} \tag{133}$$

where $K_c^{\ddagger}$ represents an equilibrium constant for the formation of an activated complex from reactants: it has units of (concentration)$^{-1}$.

The other key ingredient is to recognize that, because the activated complex and reactants are assumed to be at equilibrium, then a **Gibbs free energy of activation**, denoted by $\Delta G^{\ddagger}$, can be introduced. It is defined as the molar Gibbs free energy change for the conversion of reactants into an activated complex, each substance being in its standard state (to be discussed more fully in Block 7) at the temperature of reaction; that is,

$$\Delta G^{\ddagger} = G^{\ddagger}(\text{activated complex}) - G^{\ominus}(\text{reactants})$$

The relationship between the free energy of activation, $\Delta G^{\ddagger}$, and the **standard** or **thermodynamic equilibrium constant for the formation of an activated complex** from reactants, which we denote as $K^{\ddagger}$, is:

$$\Delta G^{\ddagger} = -RT \ln K^{\ddagger} \tag{140}$$

or in an alternative form:

$$\ln K^{\ddagger} = -\frac{\Delta G^{\ddagger}}{RT} \tag{141}$$

so that

$$K^{\ddagger} = \exp\left(-\frac{\Delta G^{\ddagger}}{RT}\right) \tag{142}$$

Notice that equation 140 is of the same form as that for 'normal' chemical reactions.

■ What are the dimensions of $K^{\ddagger}$?

▨ It is dimensionless. (If you are uncertain about the concept of a standard or thermodynamic equilibrium constant, then you should return to the discussion in Section 5.2.3 of Block 1.)

In order to be able to substitute equation 142 into equation 133, we need a relationship between $K^{\ddagger}$ and $K_c^{\ddagger}$. The relationship rests on a convention: *$K^{\ddagger}$ is related to $K_c^{\ddagger}$ by dividing each concentration in the expression for $K_c^{\ddagger}$ by a standard value of concentration*, $c^{\ominus}$. (In doing this, we are implicitly assuming ideal behaviour, as we did when discussing the standard equilibrium constant for gaseous reactions in Block 1. A more detailed and general discussion will be taken up in Block 7.)

Consider a bimolecular reaction, for example the reaction between thiosulfate ion and 1-bromopropane, which we examined in Section 2:

$$S_2O_3^{2-} + C_3H_7Br \rightleftharpoons \text{activated complex} \rightarrow \text{products} \tag{143}$$

■ Write down an expression for $K^{\ddagger}$ for this reaction.

▨ $$K^{\ddagger} = \frac{[\text{activated complex}]/c^{\ominus}}{([S_2O_3^{2-}]/c^{\ominus})([C_3H_7Br]/c^{\ominus})}$$

$$= \frac{c^{\ominus}[\text{activated complex}]}{[S_2O_3^{2-}][C_3H_7Br]} \tag{144}$$

■ Hence, what is the relationship between $K^{\ddagger}$ and $K_c^{\ddagger}$?

▨ Clearly,
$$K^{\ddagger} = c^{\ominus} K_c^{\ddagger} \tag{145}$$

because $K_c^{\ddagger} = [\text{activated complex}]/[S_2O_3^{2-}][C_3H_7Br]$. This relationship will hold for *any* solution or gas-phase *bimolecular reaction*.

Usually, $c^{\ominus}$ is chosen as 1 mol dm^{-3}, so that, according to equation 145, $K^{\ddagger}$ has the same magnitude as $K_c^{\ddagger}$, *but* is dimensionless.

Combining equation 145 with equation 142 gives:

$$K_c^{\ddagger} = \frac{1}{c^{\ominus}} \exp\left(-\frac{\Delta G^{\ddagger}}{RT}\right) \qquad (146)$$

so that substituting this equation into equation 133 gives the expression:

$$k_{\text{theory}} = \frac{v^{\ddagger}}{c^{\ominus}} \exp\left(-\frac{\Delta G^{\ddagger}}{RT}\right) \qquad (147)$$

Finally, we need to decide how to treat the quantity $v^{\ddagger}$. In our detailed discussion of a gas-phase reaction, $v^{\ddagger}$ was related to the frequency of a critical vibration along the reaction coordinate. Such molecular detail is not warranted here; it is conceptually simpler merely to make the assumption that $v^{\ddagger} = kT/h$. Hence, we arrive at a general form for the theoretical rate constant for a bimolecular reaction:

$$k_{\text{theory}} = \frac{1}{c^{\ominus}} \frac{kT}{h} \exp\left(-\frac{\Delta G^{\ddagger}}{RT}\right) \qquad (148)$$

The equation suggests that an elementary reaction can be pictured in terms of a **Gibbs free energy diagram**, as drawn schematically in Figure 35. The diagram effectively depicts the difficulty experienced by the reaction in proceeding from left to right: the larger the value of $\Delta G^{\ddagger}$, the smaller the rate constant. Notice that the abscissa has not been labelled: the diagram represents the *bulk* properties of the reaction and, as such, it represents the overall behaviour of a vast number of individual reactions, each probably occurring along a *different* reaction coordinate. The horizontal direction in the diagram is used simply to separate reactants from products.

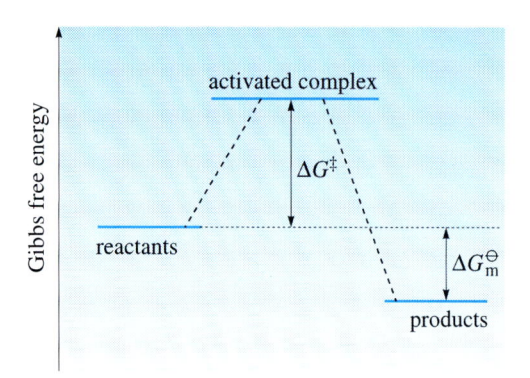

Figure 35 A Gibbs free energy diagram for an elementary reaction.

SAQ 27 The reaction in solution between thiosulfate ion and 1-bromopropane (that is, our example reaction in Section 2):

$$S_2O_3^{2-} + C_3H_7Br = C_3H_7S_2O_3^- + Br^- \qquad (5)$$

is thought to be elementary. The experimental activation energy and A-factor are 75.7 kJ mol^{-1} and 8.7×10^9 dm^3 mol^{-1} s^{-1}, respectively. What is the Gibbs free energy of activation, $\Delta G^{\ddagger}$, at 300 K?

As the answer to SAQ 27 indicates, if the rate constant for an elementary reaction is known at a particular temperature, then equation 148 can be used to determine the Gibbs free energy of activation. It follows that the larger the magnitude of this quantity, the greater the 'resistance' to the reaction proceeding from left to right.

The concepts of a Gibbs free energy of activation, as embodied in equation 148, and the schematic Gibbs free energy diagram, as in Figure 35, provide a widely used, and practical, way of describing elementary reactions in solution.

10.4 Summary of Section 10

1   The molecular potential energy curve for a diatomic molecule provides a pictorial view of the energy barrier that must be overcome if the molecule is to dissociate into atoms. The bond dissociation energy provides a good estimate of the minimum (or threshold) energy for this dissociation reaction.

2   For the gas-phase reaction F· + $H_2 \longrightarrow$ FH + H·, assuming a colinear arrangement of atoms throughout the course of reaction, the potential energy can be represented as a surface in three dimensions. Examination of this surface reveals a minimum-energy path to reaction, which passes through a transition state at the saddle point.

3   The transition state lies at the top of the energy barrier to reaction; the species at the top of this barrier is referred to as an activated complex. It is best regarded as having a definite, but entirely transitory, molecular configuration.

4   For realistic potential energy surfaces (such as for the reaction F· + $H_2 \longrightarrow$ FH + H·), the calculated minimum-energy barrier to reaction, and the experimental activation energy, are found to be of similar magnitude.

5   Any elementary reaction can be represented by a schematic energy profile, and the following relationship holds:

$$E_f - E_r = \Delta H_m^\ominus \tag{126}$$

6   Transition state theory for a bimolecular reaction pictures a reaction as proceeding by the formation of an activated complex. It is then assumed that the activated complex is in equilibrium with the reactants. Key aspects of the theory are:

- The analysis centres on the idea that the activated complex possesses a special vibrational mode that has no restoring force. When this vibration is excited, it causes the complex to fall apart into products; if you like, it is the 'Achilles heel' of the activated complex.

- Detailed evaluation of the theory can be made using the methods of statistical mechanics.

- The predictive powers of the theory are limited because of the difficulty of obtaining molecular information about the activated complex. None the less, where calculations can be made, it is far more successful than collision theory.

- The theory has the advantage that it can be formulated in more general thermodynamic terms. Equation 148 summarizes the expression for the theoretical rate constant in terms of the Gibbs free energy of activation.

$$k_{\text{theory}} = \frac{1}{c^\ominus} \frac{kT}{h} \exp\left(-\frac{\Delta G^\ddagger}{RT}\right) \tag{148}$$

OBJECTIVES FOR BLOCK 2

Now that you have completed Block 2, you should be able to do the following things:

1 Recognize valid definitions of, and use in a correct context, the terms, concepts and principles printed in bold type in the text and collected in the following Table.

List of scientific terms, concepts and principles used in Block 2

Term	Page No.	Term	Page No.
activated complex	65	molecularity	46
activation-controlled kinetics	57	molecular partition function per unit volume, Q	69
Arrhenius activation energy, E_a	36	molecular potential energy curve	60
Arrhenius A-factor, A	36	non-Arrhenius behaviour	39
Arrhenius equation	36	overall order of reaction, n	18
Arrhenius plot	37	partial order of reaction, α, β, γ, etc.	18
Beer–Lambert law	43	physical method of analysis	41
collision cross-section, σ	50	plausible rate equation	20
collision cylinder	50	potential energy profile	65
collision diameter, d	49	potential energy surface	63
collision frequency	48	psuedo-order	22
collision theory	48	pseudo-order rate constant	22
continuous flow method	44	quenching	41
contour map for an elementary reaction	64	rate constant, k_R	9
crossed molecular beams	55	rate of reaction at constant volume, J	12
differential method	20	reaction coordinate	65
diffusion-controlled reaction	58	reaction half-life, $t_{\frac{1}{2}}$	33
elementary process	45	reaction mechanism	5
empirical approach in chemical kinetics	8	reaction variable, x	15
encounter pair	57	reactive cross-section, σ_R	55
energy barrier to reaction	52	reduced mass, μ	51
entrance and exit valleys	64	saddle point or col	64
equilibrium bond length, r_e	60	solvent cage	56
experimental rate equation	9	spectroscopic dissociation energy, D_e	60
extent of reaction, ξ	13	steric factor, P	54
Eyring equation	71	stoichiometric proportions	30
flow methods	44	stoichiometric number, ν_i	12
frequency of decomposition of the activated complex, $\nu^\ddagger$	68	stopped-flow method	44
gas chromatography	42, video band 1	theoretical A-factor	53
Gibbs free energy diagram	74	thermochemical, or bond dissociation energy, D	62
Gibbs free energy of activation, $\Delta G^\ddagger$	73	threshold energy, ε_0	52
hard-sphere model	49	time-independent stoichiometry	6
harpoon mechanism	56	total collision frequency per unit volume, Z_{AB}	51
initial rate of reaction	8	total pressure measurements	42
initial rate method	22	transition state	65
integration method	20, 25	transition state theory	59
integrated rate equation	26, 29	vibrational energy level, label v	61
isolation technique	22	visible and ultraviolet spectrophotometry	43
kinetic reaction profile	7	zero-point energy	61
line-of-centres rate constant	53	zero-point internal energy of activation, $\Delta E_0^\ddagger$	66
minimum-energy path	64		

2   Describe the main aims of a typical experiment in chemical kinetics, and be able to determine the rate of change of concentration of a reactant or product species from a kinetic reaction profile. (SAQ 1)

3   Define the rate of a chemical reaction occurring at constant volume. (SAQ 2)

4   Explain why partial pressure can be used to express the concentration of a gaseous reactant, and convert the units of pressure into those of concentration. (SAQs 3, 9, 19 and 21)

5   Define the extent of a reaction, and comment on its use in defining the rate of a chemical reaction. (SAQ 4)

6   Define, and use in a correct context, the reaction variable. (SAQs 5, 10, 11, 14 and 19)

7   Given an experimental rate equation, recognize whether the concept of order has meaning for the reaction, and if so, state both the order with respect to individual species in the reaction mixture and the overall order. (SAQs 6 and 8)

8   Use the differential (or van't Hoff) method, or the method of initial rates, possibly in conjunction with an isolation technique, to establish the form of an experimental rate equation, and hence the order with respect to individual species in the reaction mixture. Indicate the advantages and disadvantages of the methods. (SAQs 8 and 18)

9   Explain why chemical reactions may have pseudo-order rate equations. (SAQ 7)

10 Explain how to select and use an integrated rate equation to determine the rate constant for a chemical reaction. (SAQs 9, 11, 12, 13 and 14)

11 Determine the length of time for a first-order reaction to progress to a certain extent. (SAQ 10)

12 Define the half-life of a chemical reaction, and indicate how its dependence on initial concentration can be used as a preliminary test for the overall order of reaction. (SAQ 15)

13 Indicate how integrated rate equations may be used in the determination of the overall order of a reaction. (SAQ 18)

14 Describe the effect of temperature on the rate of a chemical reaction, and use appropriate methods to calculate the Arrhenius activation energy and the Arrhenius A-factor from experimental data. (SAQs 16 and 18)

15 Use the Arrhenius equation in problems concerned with the change in the rate of a process with temperature. (SAQ 17)

16 Outline the main features of conventional and 'fast reaction' techniques for investigating the rates of chemical reactions, and, in particular, indicate how total pressure measurements and spectroscopic measurements are used. (SAQs 19 and 20)

17 Discuss the meaning of the term 'elementary reaction', and distinguish between the concepts of order and molecularity for a chemical reaction.

18 Outline the main features of the hard-sphere collision theory, and calculate the total collision frequency per unit volume for a bimolecular gas-phase reaction. (SAQs 21 and 22)

19 With reference to the hard-sphere collision theory for a bimolecular gas phase reaction, determine the steric factor and discuss the significance of this quantity. (SAQ 23)

20 Explain the term reactive cross-section, and indicate the types of experiment possible with a crossed molecular beam apparatus.

21 Outline a model for a bimolecular solution-phase reaction, and distinguish between activation control and diffusion control for such a reaction.

22 Use a potential energy contour map to describe the energy changes that take place as a function of the positions of atoms in an elementary gas-phase reaction. (SAQ 24)

23 Draw a schematic energy profile for an elementary reaction, and label the diagram with the enthalpy change and activation energies for the forward and reverse reactions. (SAQ 25)

24 Outline the main features of transition state theory and, given suitable data, use the Eyring equation to calculate a theoretical value for the A-factor of a bimolecular reaction. (SAQ 26)

25 Indicate how transition state theory can be developed in a form that incorporates thermodynamic parameters.

26 Given suitable kinetic data, calculate the Gibbs free energy of activation for an elementary gas- or solution-phase reaction, and plot an appropriate Gibbs free energy diagram. (SAQ 27)

SAQ ANSWERS AND COMMENTS

SAQ 1 (Objective 2)

The value of d[Br$^-$]/dt after 7 500 seconds of reaction can be found by determining the slope of the tangent to the kinetic reaction profile for Br$^-$ at this time. (Hints about the best way to draw a tangent to a curve are given in Section 3 of the AV Booklet.) You should find a value close to 1.8×10^{-6} mol dm^{-3} s^{-1}; if not, your error probably lies in not drawing a good tangent to the curve.

Clearly, Figure 1 shows that d[C$_3$H$_7$S$_2$O$_3^-$]/dt has the same value as d[Br$^-$]/dt after 7 500 seconds of reaction (and indeed throughout the reaction). This result is a direct consequence of the stoichiometry of the reaction.

The values of d[S$_2$O$_3^{2-}$]/dt and d[C$_3$H$_7$Br]/dt after 7 500 seconds of reaction are both close to -1.8×10^{-6} mol dm^{-3} s^{-1}; that is, they are *equal* to one another, and equal in magnitude, *but of opposite sign*, to d[Br$^-$]/dt. You should have been able to deduce this from the stoichiometry of the reaction, rather than by drawing tangents to the appropriate reaction profiles.

Notice that if we define the rate of reaction in terms of the rate of change of concentration of a *reactant* species, it becomes a negative quantity.

SAQ 2 (Objective 3)

The 'recipe' for writing down the rate of reaction under constant-volume conditions in terms of the rate of change of concentration of either a reactant or product species is given by equation 11. For reaction 14

$$J = -\frac{1}{2}\frac{d[H_2]}{dt} = -\frac{d[O_2]}{dt} = \frac{1}{2}\frac{d[H_2O]}{dt}$$

For reaction 15

$$J = -\frac{1}{2}\frac{d[NO]}{dt} = -\frac{1}{2}\frac{d[H_2]}{dt} = \frac{d[N_2]}{dt} = \frac{1}{2}\frac{d[H_2O]}{dt}$$

For reaction 16

$$J = -\frac{d[BrO_3^-]}{dt} = -\frac{1}{5}\frac{d[Br^-]}{dt} = -\frac{1}{6}\frac{d[H^+]}{dt} = \frac{1}{3}\frac{d[Br_2]}{dt} = \frac{1}{3}\frac{d[H_2O]}{dt}$$

SAQ 3 (Objective 4)

Assuming the ideal gas law is obeyed, then for a reactant A, according to equation 12

$$p(A) = \frac{n_A RT}{V}, \text{ or } \frac{n_A}{V} = \frac{p(A)}{RT}$$

Now, $p(A)$ is 1.280×10^4 Pa (that is, 1.280×10^4 J m^{-3}), so

$$\frac{n_A}{V} = \frac{(1.280 \times 10^4 \text{ J m}^{-3})}{(8.314 \text{ J K}^{-1} \text{ mol}^{-1}) \times (350 \text{ K})}$$

$$= 4.40 \text{ mol m}^{-3}$$

$$= 4.40 \text{ mol} \times (10 \text{ dm})^{-3}$$

$$= 4.40 \times 10^{-3} \text{ mol dm}^{-3}$$

SAQ 4 (Objective 5)

The formal definition of extent of reaction is given by equation 17. Hence, using the data in Table 2, in terms of N_2O_5,

$$\xi = \frac{n_{N_2O_5} - 3.00 \times 10^{-3} \text{ mol}}{-2}$$

in terms of NO_2,

$$\xi = \frac{n_{NO_2} - 0 \text{ mol}}{4}$$

and in terms of O_2,

$$\xi = \frac{n_{O_2} - 0 \text{ mol}}{1}$$

Any one of these three expressions can be used to calculate the extent of reaction, since at a given time this quantity will be the same, irrespective of whether it is calculated in terms of the amount of reactant consumed, or the amounts of products formed. Table 14 summarizes the calculations.

Table 14 The thermal decomposition of N_2O_5 at 328.1 K: calculation of the extent of reaction.

$\dfrac{\text{time}}{\text{s}}$	$\dfrac{n_{N_2O_5}}{10^{-3} \text{ mol}}$	$\dfrac{n_{NO_2}}{10^{-3} \text{ mol}}$	$\dfrac{n_{O_2}}{10^{-3} \text{ mol}}$	$\dfrac{\xi}{10^{-3} \text{ mol}}$
0	3.00	0	0	0
580	1.22	3.56	0.89	0.89
1 060	0.59	4.82	1.205	1.205
∞	0	6.00	1.50	1.50

SAQ 5 (Objective 6)

Let the initial concentration of potassium iodide be represented by the quantity $[KI]_0$. According to the definition of the reaction variable given in equation 20, we can write (remembering that $\nu_{KI} = -3$)

$$x = \frac{[KI] - [KI]_0}{-3}$$

Thus, the concentration of potassium iodide, $[KI]$, at any time in the reaction can be calculated from:

$$[KI] = [KI]_0 - 3x$$

SAQ 6 (Objective 7)

Reaction (a) is second order with respect to NO, first order with respect to O_2, and third order overall.

Reaction (b) is first order with respect to CO, and order $\frac{3}{2}$ with respect to Cl_2. The overall order is $\frac{5}{2}$.

Reaction (c) is first order with respect to both Fe^{2+} and H_2O_2, and second order overall. (Notice that the rate equation is independent of the H^+ concentration, so that the partial order with respect to this species is *zero*: any quantity raised to the power zero is unity.)

Reaction (d) is second order with respect to O_3, order -1 with respect to O_2, and first order overall. Notice that in this case, O_2 is a *product* species.

Reaction (e) is not of the general form given by equation 24, so that neither partial orders with respect to individual species in the reaction mixture nor the overall order can be defined.

SAQ 7 (Objective 9)

Because water is the solvent in the reaction, it will be present in large excess compared with the methyl ethanoate. Therefore, one possibility is that the experimental rate equation is pseudo first order. In such circumstances, the rate constant k_R will be a pseudo-first-order rate constant, whose value depends on the concentration of water.

(In fact, experimental investigation confirms this possibility, and it is found that $k_R = k_R'[H_2O]$, where k_R' is the true *second-order* rate constant.)

SAQ 8 (Objectives 7 and 8)

According to equation 11, the rate of reaction at constant volume can be expressed as:

$$J = -\frac{d[IO_3^-]}{dt} = -\frac{1}{5}\frac{d[I^-]}{dt} = -\frac{1}{6}\frac{d[H^+]}{dt} = \frac{1}{3}\frac{d[I_2]}{dt} = \frac{1}{3}\frac{d[H_2O]}{dt}$$

A possible rate equation, by analogy with equation 24, is

$$J = k_R[IO_3^-]^\alpha[I^-]^\beta[H^+]^\gamma$$

Since Table 5 gives initial rate data, we can write

$$J_0 = k_R[IO_3^-]_0^\alpha [I^-]_0^\beta [H^+]_0^\gamma$$

In the first three rows of the Table, both $[H^+]_0$ and $[I^-]_0$ are constant. Therefore

$$J_0 = k_R'[IO_3^-]_0^\alpha$$

where

$$k_R' = k_R[I^-]_0^\beta [H^+]_0^\gamma$$

By inspection, it is clear that J_0 is directly proportional to $[IO_3^-]_0$, so that $\alpha = 1$.

In rows 3 and 5, both $[H^+]_0$ and $[IO_3^-]_0$ have constant values, so that

$$J_0 = k_R''[I^-]_0^\beta$$

where

$$k_R'' = k_R[IO_3^-]_0^\alpha [H^+]_0^\gamma$$

Clearly, doubling the value of $[I^-]_0$ causes the initial rate to increase by a factor of 4, and hence $\beta = 2$. This conclusion is supported by the data in rows 4 and 6, which show that with $[H^+]_0$ and $[IO_3^-]_0$ again constant, the value of J_0 increases by a factor of approximately 16 when the value of $[I^-]_0$ is quadrupled.

In rows 3 and 4, both $[I^-]_0$ and $[IO_3^-]_0$ are constant. Therefore

$$J_0 = k_R'''[H^+]_0^\gamma$$

where

$$k_R''' = k_R[IO_3^-]_0^\alpha [I^-]_0^\beta$$

By inspection, doubling the value of $[H^+]_0$ causes the initial rate to increase by a factor of 4, and hence $\gamma = 2$.

The rate equation is thus

$$J = k_R[IO_3^-][I^-]^2[H^+]^2$$

although in terms of the information given, we can be certain only that this equation is valid for the *initial* stages of reaction. The reaction is first order with respect to IO_3^-, second order with respect to I^- and second order with respect to H^+; hence, overall, it is fifth order.

A value of the rate constant at 298 K can be found by calculating the average value of the quantity $J_0/[IO_3^-]_0[I^-]_0^2[H^+]_0^2$ using the data in Table 5. You should find $k_R = 3.0 \times 10^8 \, dm^{12} \, mol^{-4} \, s^{-1}$. *Note very carefully the units of this rate constant.*

SAQ 9 (Objectives 4 and 10)

Figure 8 shows a plot of $\ln \{p(N_2O_5)/Pa\}$ versus time; thus, the partial pressure of N_2O_5 has been taken to be proportional to its concentration. If we assume ideal gas behaviour, then according to equation 13 in Section 3:

$$p(N_2O_5) = [N_2O_5]RT$$

so in terms of partial pressure, equation 46 can be written as

$$\ln(p_0(N_2O_5)/RT) - \ln(p(N_2O_5)/RT) = 2k_R t$$

If the logarithms are expanded, then

$$[\ln(p_0(N_2O_5)) + \ln(1/RT)] - [\ln(p(N_2O_5)) + \ln(1/RT)] = 2k_R t$$

and so the equation can be simplified

$$\ln(p_0(N_2O_5)) - \ln(p(N_2O_5)) = 2k_R t$$

or, on rearrangement

$$\ln(p(N_2O_5)) = -2k_R t + \ln(p_0(N_2O_5))$$

Thus, a plot of $\ln(p(N_2O_5)/Pa)$ versus time will give a straight line of the same slope as a plot of $\ln([N_2O_5]/mol\,dm^{-3})$ versus time. The slope will be equal to $-2k_R$.

From the Figure

$$\text{slope} = \frac{7.95 - 10.24}{1\,800\,s - 300\,s} = -1.527 \times 10^{-3}\,s^{-1}$$

Hence, $-2k_R = -1.527 \times 10^{-3}\,s^{-1}$, so $k_R = 7.635 \times 10^{-4}\,s^{-1}$ at 328.1 K.

Notice the 'factor of two' that enters into the calculation of k_R from the slope; this is a direct consequence of the stoichiometry of the reaction.

SAQ 10 (Objectives 6 and 11)

According to equation 50

$$[N_2O_5] = [N_2O_5]_0 \exp(-2k_R t)$$

where $[N_2O_5]_0$ is the initial concentration of N_2O_5. When 90% has decomposed, $[N_2O_5] = 0.1[N_2O_5]_0$. Hence, using the value of $k_R = 7.635 \times 10^{-4}\,s^{-1}$ from SAQ 9, it follows that:

$$0.1[N_2O_5]_0 = [N_2O_5]_0 \exp(-2 \times 7.635 \times 10^{-4}\,s^{-1} \times t)$$

or, on dividing through by $[N_2O_5]_0$

$$0.1 = \exp(-1.527 \times 10^{-3}\,s^{-1} \times t)$$

and so, taking natural logarithms,

$$-2.303 = -1.527 \times 10^{-3}\,s^{-1} \times t$$

Hence, $t = 1\,508\,s$.

(Remember that if $q = e^x = \exp(x)$, then x is *defined* to be the natural logarithm of q; that is, $\ln q = x$. The final step in the calculation above is just a specific example of this general result. If you are unsure about this, it would be a good plan to consult Section 1 of the AV Booklet.)

Notice an important feature of this calculation: the length of time it takes for 90% (or any other percentage) of the initial concentration of a reactant in a *first-order reaction* to decompose is independent of its initial concentration. Hence, you could have answered this question – although not as accurately – by reading the time from Figure 4. (Try this for yourself.)

Equation 50 can be expressed in terms of the concentration of the *product* NO_2, as follows. In terms of NO_2 the reaction variable is

$$x = \frac{[NO_2]}{4}; \text{ that is, } [NO_2] = 4x$$

and in terms of N_2O_5,

$$x = \frac{[N_2O_5] - [N_2O_5]_0}{-2}; \text{ that is, } [N_2O_5] = [N_2O_5]_0 - 2x.$$

Hence,

$$[N_2O_5] = [N_2O_5]_0 - \tfrac{1}{2}[NO_2]$$

and so substituting into equation 50 gives

$$[N_2O_5]_0 - \tfrac{1}{2}[NO_2] = [N_2O_5]_0 \exp(-2k_R t)$$

This equation can be rearranged as follows

$$\tfrac{1}{2}[NO_2] = [N_2O_5]_0 - [N_2O_5]_0 \exp(-2k_R t)$$

or

$$[NO_2] = 2[N_2O_5]_0 \{1 - \exp(-2k_R t)\}$$

After 1 508 s of reaction (according to the answer to the first part of the question), the quantity $\exp(-2k_R t)$ is equal to 0.1, so

$$[NO_2] = 2 \times 0.01 \text{ mol dm}^{-3} \times (1 - 0.1)$$

$$[NO_2] = 0.018 \text{ mol dm}^{-3}$$

Of course, the same answer could have been arrived at by just inspecting the stoichiometry of the reaction. However, the formal derivation illustrates the usefulness of the reaction variable, particularly when dealing with reactions of fairly complex stoichiometry.

SAQ 11 (Objectives 6 and 10)

The reaction between 1,2-dibromoethane and potassium iodide is:

$$C_2H_4Br_2 + 3KI = KI_3 + 2KBr + C_2H_4 \qquad (22)$$

If A is taken to represent $C_2H_4Br_2$ and B to represent KI, then $a = 1$ and $b = 3$, so according to equation 60 in Table 7:

$$\ln\left(\frac{[C_2H_4Br_2]}{[KI]}\right) = (3[C_2H_4Br_2]_0 - [KI]_0)k_R t + \ln\left(\frac{[C_2H_4Br_2]_0}{[KI]_0}\right)$$

But according to Table 6, $[C_2H_4Br_2] = [C_2H_4Br_2]_0 - x$, and $[KI] = [KI]_0 - 3x$, so

$$\ln\left\{\frac{[C_2H_4Br_2]_0 - x}{([KI]_0 - 3x)}\right\} = (3[C_2H_4Br_2]_0 - [KI]_0)k_R t + \ln\left\{\frac{[C_2H_4Br_2]_0}{[KI]_0}\right\}$$

SAQ 12 (Objective 10)

The most accurate method for determining the rate constant of a reaction is to use the appropriate integrated rate equation. The information in Figure 1 (and Table 1) is for experimental conditions in which the reactants had different initial concentrations (the thiosulfate ion was in excess). The appropriate integrated rate equation is thus given by equation 59 in Table 7; that is,

$$\ln\left(\frac{[S_2O_3^{2-}]}{[C_3H_7Br]}\right) = ([S_2O_3^{2-}]_0 - [C_3H_7Br]_0)k_R t + \ln\left(\frac{[S_2O_3^{2-}]_0}{[C_3H_7Br]_0}\right)$$

A plot of $\ln([S_2O_3^{2-}]/[C_3H_7Br])$ versus time should thus be a straight line with slope $([S_2O_3^{2-}]_0 - [C_3H_7Br]_0)k_R$, from which a value of the rate constant can be calculated.

SAQ 13 (Objective 10)

Table 15 gives the data for the second-order plot, which is shown in Figure 36.

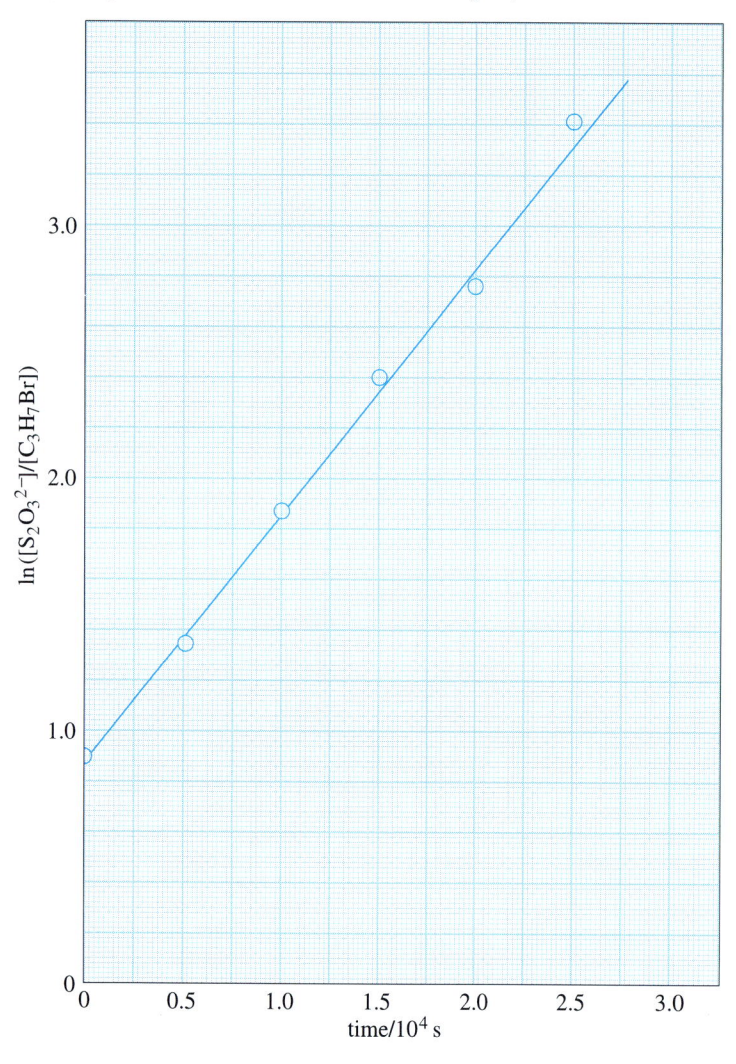

Figure 36 A second-order plot for the reaction between thiosulfate ion and 1-bromopropane at 310.7 K.

Table 15 Determination of the second-order rate constant for the reaction between thiosulfate ion and 1-bromopropane at 310.7 K.

$\dfrac{\text{time}}{10^4 \text{ s}}$	$\dfrac{[S_2O_3^{2-}]}{\text{mol dm}^{-3}}$	$\dfrac{[C_3H_7Br]}{\text{mol dm}^{-3}}$	$\ln\left(\dfrac{[S_2O_3^{2-}]}{[C_3H_7Br]}\right)$
0	0.100	0.041	0.89
0.5	0.081	0.021	1.35
1.0	0.071	0.011	1.86
1.5	0.066	0.006	2.40
2.0	0.063	0.004	2.76
2.5	0.061	0.002	3.42

From Figure 36

$$\text{slope} = \frac{3.29 - 1.36}{2.5 \times 10^4 \text{ s} - 0.5 \times 10^4 \text{ s}} = 9.65 \times 10^{-5} \text{ s}^{-1}$$

The initial concentrations are

$[S_2O_3^{2-}]_0 = 0.100 \text{ mol dm}^{-3}$ and $[C_3H_7Br]_0 = 0.041 \text{ mol dm}^{-3}$

so, $(0.100 \text{ mol dm}^{-3} - 0.041 \text{ mol dm}^{-3})k_R = 9.65 \times 10^{-5} \text{ s}^{-1}$

and hence

$k_R = 1.64 \times 10^{-3} \text{ dm}^3 \text{ mol}^{-1} \text{ s}^{-1}$

This value compares favourably with that found by taking the ratio $J/[S_2O_3^{2-}][C_3H_7Br]$ as carried out in Table 1.

Notice that the experimental points are scattered, particularly at *long* times; this reflects the fact that the relative uncertainty in measuring concentrations increases as the reaction nears completion. The question arises as to how to draw a good straight line through the experimental points. In this Course we suggest that you simply rely on the judgement of your eye, which more often than not gives quite acceptable results. But there are other methods, which you will have the opportunity to learn about at the Residential School.

SAQ 14 (Objectives 6 and 10)

The rate constant is best determined using an appropriate integrated rate equation. Since, experimentally, the reaction is found to be second order, and the initial conditions are such that the peroxodisulfate ion is in slight excess, then the integrated rate equation required is that given by equation 60 in Table 7. To use this equation it is necessary to determine how the concentrations of both peroxodisulfate ion and iodide ion change during the course of the reaction. However, you are told that the reaction is followed by monitoring the change in concentration of iodine. To determine changes in reactant concentrations it is best to make use of the reaction variable: Table 16 shows how to do this.

Table 16 Using the reaction variable for the reaction in equation 66.

Substance, Y	v_Y	$[Y]_0$	$[Y]$
$S_2O_8^{2-}$	-1	$[S_2O_8^{2-}]_0$	$[S_2O_8^{2-}]_0 - x$
I^-	-2	$[I^-]_0$	$[I^-]_0 - 2x$
SO_4^{2-}	$+2$	0	$2x$
I_2	$+1$	0	x

The Table indicates that the concentration of iodine at any time during the reaction is equal to the reaction variable, so that the concentrations of reactants at any time during the reaction can be calculated as long as their initial concentrations are known. A suitable second-order plot based on equation 60 can then be made in order to determine the rate constant.

SAQ 15 (Objective 12)

Clearly, the reaction cannot be first order, because the reaction half-life depends on the initial concentration of reactants.

If the reaction is second order, then:

$$-\frac{d[IF_5]}{dt} = -\frac{d[F_2]}{dt} = k_R[IF_5][F_2]$$

But the initial concentrations of the reactants are the same; that is, the reactants are in stoichiometric proportions, so that throughout the course of the reaction $[IF_5] = [F_2]$. Thus, the rate equation can be written as

$$-\frac{d[F_2]}{dt} = k_R[F_2]^2$$

and so according to equation 77 (with $a = 1$)

$$t_{\frac{1}{2}} = \frac{1}{k_R [F_2]_0}$$

Hence, to confirm that the reaction is second order, we must show that the product $t_{\frac{1}{2}}[F_2]_0$ is a constant. Taking each row in turn in Table 8, we find the product equals 86.9, 87.0 and 86.6 mol dm^{-3} s; that is, within experimental error it is a constant. The reaction is therefore second order.

SAQ 16 (Objective 14)

According to equation 82, the slope of a straight line obtained by plotting log k_R versus $1/T$ is

$$\text{slope} = -\frac{E_a}{2.303 R}$$

Thus, as a first step in determining E_a, we need to determine the slope of the straight line in Figure 14.

Notice how the reciprocal temperature axis has been labelled. If, say, 10^3 K/$T = 2.7$, then $1/T = 2.7 \times 10^{-3}$ K^{-1}. (You may recall that you have already met this way of labelling a reciprocal temperature axis in Figure 1 of Block 1.)

Hence, from Figure 14,

$$\text{slope} = \frac{-3.93 - (-1.20)}{(3.2 \times 10^{-3} \text{ K}^{-1}) - (2.7 \times 10^{-3} \text{ K}^{-1})}$$

$$= \frac{-2.73}{0.5 \times 10^{-3} \text{ K}^{-1}}$$

$$= -5.46 \times 10^3 \text{ K}$$

and so it follows that

$$-5.46 \times 10^3 \text{ K} = -\frac{E_a}{2.303 R}$$

The activation energy is then calculated as follows

$$E_a = -(-5.46 \times 10^3 \text{ K} \times 2.303 \times 8.314 \text{ J K}^{-1} \text{ mol}^{-1})$$

$$= 104.5 \times 10^3 \text{ J mol}^{-1}, \text{ or } 104.5 \text{ kJ mol}^{-1}$$

SAQ 17 (Objective 15)

At 285 K:

$$\text{rate}(285 \text{ K}) = \text{constant} \times \exp(-E_a/(R \times 285 \text{ K}))$$

At a later time, let the temperature in the forest be T:

$$\text{rate}(T) = \text{constant} \times \exp(-E_a/RT)$$

But we are told that

$$\frac{\text{rate}(T)}{\text{rate}(285 \text{ K})} = 2$$

so that

$$\frac{\text{rate}(T)}{\text{rate}(285 \text{ K})} = 2 = \frac{\exp(-E_a/RT)}{\exp\{-E_a/(R \times 285 \text{ K})\}}$$

or, on simplifying,

$$2 = \exp\left\{\left(\frac{-E_a}{RT}\right) - \left(\frac{-E_a}{(R \times 285\text{ K})}\right)\right\}$$

$$= \exp\left\{\frac{-E_a}{R}\left(\frac{1}{T} - \frac{1}{285\text{ K}}\right)\right\}$$

Now take natural logarithms of both sides of the equation

$$\ln 2 = -\frac{E_a}{R}\left(\frac{1}{T} - \frac{1}{285\text{ K}}\right)$$

Hence, on substituting numerical values

$$0.693 = -\frac{51.3 \times 10^3 \text{ J mol}^{-1}}{8.314 \text{ J K}^{-1} \text{ mol}^{-1}}\left(\frac{1}{T} - \frac{1}{285\text{ K}}\right)$$

$$= -6.170 \times 10^3 \text{ K}\left(\frac{1}{T} - \frac{1}{285\text{ K}}\right)$$

Rearranging this expression gives

$$\left(\frac{1}{T} - \frac{1}{285\text{ K}}\right) = -1.123 \times 10^{-4} \text{ K}^{-1}$$

So

$$\frac{1}{T} = (-0.1123 + 3.509) \times 10^{-3} \text{ K}^{-1}$$

or

$$T = \frac{1}{3.3967 \times 10^{-3} \text{ K}^{-1}}$$

which gives a final answer

$$T = 294.4 \text{ K}$$

The increase in temperature was therefore (294.4 K − 285 K); that is, 9.4 K. Notice that the process roughly obeys the 'chemical rule of thumb' mentioned at the beginning of Section 6.

SAQ 18 (Objectives 8, 13 and 14)

The strategy is as follows. *All* that is known is the stoichiometry of the reaction, and so the first step is to propose a plausible rate equation. A reasonable suggestion would be

$$J = k_R[CH_3I]^\alpha[CH_3CH_2O^-]^\beta$$

To determine the partial orders of reaction, α and β, the best approach is to use the *differential method*, and since *two* reactants are involved, an *isolation technique* would be convenient. For example, an experimental run in which the concentration of iodomethane is in large excess (that is, $[CH_3I]_0 \gg [CH_3CH_2O^-]_0$) can be used. In this case, the rate equation is effectively reduced to:

$$J = k_R'[CH_3CH_2O^-]^\beta$$

where $k_R' = k_R[CH_3I]_0^\alpha$. If a *kinetic reaction profile* in terms of the change in concentration of ethoxide ion with time is obtained, then this can be used to determine J as a function of the concentration of ethoxide ion. Hence, a plot of log J versus log $[CH_3CH_2O^-]$ should be a straight line with a slope from which the value of β can be determined. In a similar manner, the partial order with respect to iodomethane can be measured under conditions such that $[CH_3CH_2O^-]_0 \gg [CH_3I]_0$.

Once the order of reaction has been established, the rate constant can be determined using an *appropriate integrated rate equation*. Which equation to use will depend on the form of the rate equation and on the initial conditions; for example, whether one reactant is in large excess, or whether the reactants are mixed in stoichiometric proportions.

SAQ 19 (Objectives 4, 6 and 16)

Assuming ideal gas behaviour, then the partial pressure of any species – a reactant or a product in the reaction mixture – can be expressed as (equation 13)

$$p(Y) = [Y]RT$$

or

$$[Y] = \frac{p(Y)}{RT}$$

The extent of reaction per unit volume, x, (Section 3.1) is defined by

$$x = \frac{[Y] - [Y]_0}{\nu_Y}$$

and so it follows that

$$x = \frac{\{p(Y)/RT\} - \{p_0(Y)/RT\}}{\nu_Y}$$

Multiplying both sides of this equation by the term RT gives

$$xRT = \frac{p(Y) - p_0(Y)}{\nu_Y}$$

or

$$x' = \frac{p(Y) - p_0(Y)}{\nu_Y}$$

where $x' = xRT$.

For the decomposition reaction

$$2N_2O_5(g) = 4NO_2(g) + O_2(g)$$

it is useful to summarize the quantities of interest in tabular form, as in Table 17.

Table 17 Using a 'modified reaction variable' for the decomposition of N_2O_5.

Substance, Y	ν_Y	$p_0(Y)$	$p(Y)$
N_2O_5	-2	$p_0(N_2O_5)$	$p_0(N_2O_5) - 2x'$
NO_2	4	0	$4x'$
O_2	1	0	x'

Applying Dalton's law of partial pressures

$$p_{tot} = p(N_2O_5) + p(NO_2) + p(O_2)$$

or

$$\begin{aligned} p_{tot} &= p_0(N_2O_5) - 2x' + 4x' + x' \\ &= p_0(N_2O_5) + 3x' \end{aligned}$$

This expression allows the 'modified reaction variable' to be expressed in terms of the initial pressure of N_2O_5 and the total pressure; that is,

$$3x' = p_{tot} - p_0(N_2O_5)$$

or

$$x' = \tfrac{1}{3}\{p_{tot} - p_0(N_2O_5)\}$$

At any time in the reaction, the partial pressure of N_2O_5 can be expressed as (Table 17)

$$p(N_2O_5) = p_0(N_2O_5) - 2x'$$
$$= p_0(N_2O_5) - \tfrac{2}{3}\{p_{tot} - p_0(N_2O_5)\}$$
$$= \tfrac{1}{3}\{5p_0(N_2O_5) - 2p_{tot}\}$$

The 'modified reaction variable' can be very useful when considering the relationship between the partial pressures of reactants and products in gas-phase reactions that have relatively complex stoichiometries.

SAQ 20 (Objective 16)

The reaction is second order, so that

$$-\frac{d[Fe^{2+}]}{dt} = -\frac{d[(Co(C_2O_4)_3)^{3-}]}{dt} = k_R[Fe^{2+}][(Co(C_2O_4)_3)^{3-}]$$

The initial concentrations of reactants are in stoichiometric proportions, so that throughout the course of the reaction, $[Fe^{2+}] = [(Co(C_2O_4)_3)^{3-}]$. Hence, the rate equation can be written as:

$$-\frac{d[Fe^{2+}]}{dt} = k_R[Fe^{2+}]^2$$

and the half-life according to equation 77 is

$$t_{\frac{1}{2}} = \frac{1}{k_R[Fe^{2+}]_0}$$

Hence, at 318.5 K

$$t_{\frac{1}{2}} = \frac{1}{(9.8 \times 10^2 \text{ dm}^3 \text{ mol}^{-1} \text{ s}^{-1}) \times (1.0 \times 10^{-3} \text{ mol dm}^{-3})}$$
$$= 1.02 \text{ s}$$

In order for conventional methods to be used to investigate the reaction, the half-life needs to be longer than 10 s. The way to achieve this is to reduce the initial concentrations of reactants by at least an order of magnitude – but, of course, sensitive analytical techniques must then be available for monitoring the reaction.

SAQ 21 (Objectives 4 and 18)

If a gas A is assumed to behave ideally, then, according to equation 12 in Section 3

$$p(A) = \frac{n_A RT}{V}$$

If the gas in a fixed volume V contains N_A molecules or atoms, then the number of moles of A present is $n_A = N_A/L$, where L is the Avogadro constant ($L = 6.022 \times 10^{23}$ mol^{-1}). Hence,

$$p(A) = \frac{N_A RT}{LV}$$

so that

$$\text{molecules of A per unit volume} = \frac{N_A}{V} = \frac{p(A)L}{RT}$$

Thus, for a gas of partial pressure 1.0×10^5 Pa (remember 1 Pa = 1 J m^{-3}) at 400 K

$$\frac{N_A}{V} = \frac{1.0 \times 10^5 \text{ J m}^{-3} \times 6.022 \times 10^{23} \text{ mol}^{-1}}{8.314 \text{ J K}^{-1} \text{ mol}^{-1} \times 400 \text{ K}}$$
$$= 1.811 \times 10^{25} \text{ (molecules) m}^{-3}$$

SAQ 22 (Objective 18)

The total collision frequency is calculated from equation 106, combined with the definitions in equations 107 and 108, and taking $\sigma = \pi d^2$. It is best to approach the calculation in stages.

The mass of one molecule of ethene, C_2H_4, is

$$m(C_2H_4) = \frac{28.054 \times 10^{-3} \text{ kg mol}^{-1}}{6.022 \times 10^{23} \text{ mol}^{-1}}$$
$$= 4.659 \times 10^{-26} \text{ kg}$$

The mass of one molecule of buta-1,3-diene, C_4H_6, is

$$m(C_4H_6) = \frac{54.092 \times 10^{-3} \text{ kg mol}^{-1}}{6.022 \times 10^{23} \text{ mol}^{-1}}$$
$$= 8.982 \times 10^{-26} \text{ kg}$$

Hence, the reduced mass can be calculated from equation 108

$$\mu = \frac{(4.659 \times 10^{-26} \text{ kg}) \times (8.982 \times 10^{-26} \text{ kg})}{(4.659 \times 10^{-26} \text{ kg}) + (8.982 \times 10^{-26} \text{ kg})}$$
$$= 3.068 \times 10^{-26} \text{ kg}$$

According to equation 107

$$\bar{u}_{AB} = \left(\frac{8 \times 1.3807 \times 10^{-23} \text{ J K}^{-1} \times 400 \text{ K}}{\pi \times 3.068 \times 10^{-26} \text{ kg}} \right)^{1/2}$$

$$\bar{u}_{AB} = 6.770 \times 10^2 \text{ m s}^{-1} \text{ (since J = kg m}^2 \text{ s}^{-2})$$

The collision cross-section $\sigma = \pi \times (500 \times 10^{-12} \text{ m})^2 = 7.854 \times 10^{-19} \text{ m}^2$.

The partial pressure of each gas at 400 K is 10^5 Pa, so according to the answer to SAQ 21, the number of molecules per unit volume of each gas is 1.811×10^{25} m^{-3}.

Pulling the various strands of the calculation together

$$Z_{AB} = \sigma \bar{u}_{AB} \frac{N_A N_B}{V^2}$$
$$= (7.854 \times 10^{-19} \text{ m}^2) \times (6.770 \times 10^2 \text{ m s}^{-1}) \times (1.811 \times 10^{25} \text{ m}^{-3})^2$$
$$= 1.744 \times 10^{35} \text{ m}^{-3} \text{ s}^{-1}$$

Thus, on average, there are some 10^{35} collisions per second in a volume of 1 m^3 containing of the order of 10^{25} molecules.

SAQ 23 (Objective 19)

First, a value for the theoretical A-factor must be calculated. This is given by equation 118:

$$A_{\text{theory}} = L\pi d^2 \left(\frac{8kT}{\pi \mu} \right)^{1/2} = L\pi d^2 \bar{u}_{AB}$$

This theoretical A-factor is temperature dependent. As a compromise, the calculation is carried out at the mid-point of the experimental temperature range, namely 2 700 K. (In fact, the calculation is hardly altered by using 2 400 K or 3 000 K.)

The average relative speed of CO and O_2 at 2 700 K is given in the question; therefore:

$$A_{theory} = (6.022 \times 10^{23}\,mol^{-1}) \times \pi \times (365 \times 10^{-12}\,m)^2 \times (1.957 \times 10^3\,m\,s^{-1})$$
$$= 4.93 \times 10^8\,m^3\,mol^{-1}\,s^{-1}$$

or, in more conventional units,

$$A_{theory} = 4.93 \times 10^{11}\,dm^3\,mol^{-1}\,s^{-1}$$

The experimental A-factor is $3.5 \times 10^9\,dm^3\,mol^{-1}\,s^{-1}$, so:

$$\text{steric factor, } P = \frac{3.5 \times 10^9\,dm^3\,mol^{-1}\,s^{-1}}{4.93 \times 10^{11}\,dm^3\,mol^{-1}\,s^{-1}}$$
$$= 0.007$$

It is difficult to see how a term supposedly incorporating only geometrical factors should be so small for such an apparently simple reaction.

SAQ 24 (Objective 22)

At the position labelled Z on the contour map, the atoms F·, H· and H´· are well separated and are essentially independent of one another.

A cross-section along A········A´ has r(H---H´) constant, and so the resulting potential energy curve will be closely related to that of the HF molecule, but only identical with it when r(H---H´) is extremely large.

A cross-section along B········B´ has r(F---H) constant, and so the resulting potential energy curve will be closely related to that of the H_2 molecule, but only identical with it when r(F---H) is extremely large.

Notice that the contours indicate that the 'HF valley' is steeper than the 'H_2 valley', as we suggested in the schematic sketch in Figure 28.

SAQ 25 (Objective 23)

A schematic energy profile for an endothermic elementary reaction is drawn in Figure 37.

As the diagram indicates, the potential energy barrier for the forward reaction, E_f, must be *at least* as high as the enthalpy change for the reaction. In other words, the enthalpy change sets a *lower limit* to the activation energy. In terms of equation 126, the smallest possible value of E_r is zero, and so $E_f \geq \Delta H_m^\ominus$.

This contrasts with the case of an exothermic reaction, where a knowledge of $\Delta H_m^\ominus$ alone tells us nothing about the activation energy for the forward reaction.

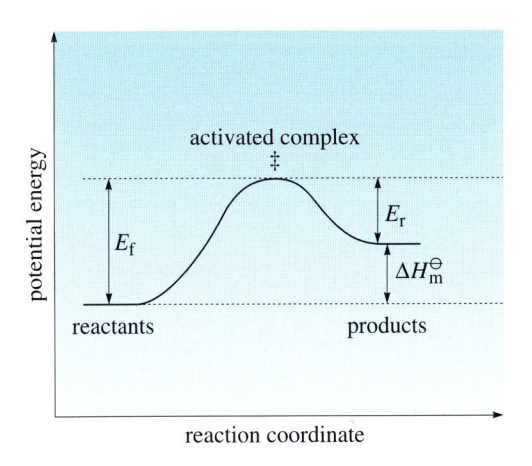

Figure 37 A schematic energy profile for an endothermic elementary reaction.

SAQ 26 (Objective 24)

According to equation 139

$$A_{\text{theory}} = \frac{kT}{h} L \frac{Q^{\ddagger\prime}}{Q_{\text{reactants}}}$$

If at 400 K the ratio $Q^{\ddagger\prime}/Q_{\text{reactants}}$ has the value 10^{-33} m^3, then

$$A_{\text{theory}} = \left(\frac{1.3807 \times 10^{-23} \text{ J K}^{-1} \times 400 \text{ K}}{6.6262 \times 10^{-34} \text{ J s}}\right) \times (6.022 \times 10^{23} \text{ mol}^{-1}) \times 10^{-33} \text{ m}^3$$

$$= 5.02 \times 10^3 \text{ m}^3 \text{ mol}^{-1} \text{ s}^{-1}$$

$$= 5.02 \times 10^3 (10 \text{ dm})^3 \text{ mol}^{-1} \text{ s}^{-1}$$

$$= 5.02 \times 10^6 \text{ dm}^3 \text{ mol}^{-1} \text{ s}^{-1}$$

This theoretical value is significantly smaller than that expected from a hard-sphere collision theory calculation: typically, values of around 10^{11} dm^3 mol^{-1} s^{-1} are predicted.

Compare the value for the last reaction in Table 9, the reaction between ethene and buta-1,3-diene (a 'complex' reaction – at least to the physical chemist), which has an experimental A-factor of 3.2×10^7 dm^3 mol^{-1} s^{-1}. This is a value within an order of magnitude of the transition state theory estimate, but vastly different from that predicted by collision theory (3.2×10^{11} dm^3 mol^{-1} s^{-1}).

This example demonstrates clearly the ability of transition state theory to include 'molecular complexity' in the estimation of kinetic parameters.

SAQ 27 (Objective 26)

As a first step, the bimolecular second-order rate constant at 300 K must be determined from the Arrhenius equation

$$k_R = (8.7 \times 10^9 \text{ dm}^3 \text{ mol}^{-1} \text{ s}^{-1}) \exp\left(\frac{-75.7 \times 10^3 \text{ J mol}^{-1}}{8.314 \text{ J K}^{-1} \text{ mol}^{-1} \times 300 \text{ K}}\right)$$

$$= 5.735 \times 10^{-4} \text{ dm}^3 \text{ mol}^{-1} \text{ s}^{-1}$$

According to equation 148

$$k_{\text{theory}} = \frac{1}{c^{\ominus}} \frac{kT}{h} \exp\left(-\frac{\Delta G^{\ddagger}}{RT}\right)$$

Thus, if we equate k_{theory} with the experimental rate constant, k_R, and rearrange the equation, then

$$\exp(-\Delta G^{\ddagger}/RT) = \frac{c^{\ominus} k_R h}{kT}$$

The next step is to take natural logarithms:

$$\Delta G^{\ddagger} = -RT \ln\left(\frac{c^{\ominus} k_R h}{kT}\right)$$

$$= -(8.314 \text{ J K}^{-1} \text{ mol}^{-1}) \times (300 \text{ K}) \times \ln\left(\frac{1 \text{ mol dm}^{-3} \times 5.735 \times 10^{-4} \text{ dm}^3 \text{ mol}^{-1} \text{ s}^{-1} \times 6.6262 \times 10^{-34} \text{ J s}}{1.3807 \times 10^{-23} \text{ J K}^{-1} \times 300 \text{ K}}\right)$$

$$= 92.1 \text{ kJ mol}^{-1}$$

Note carefully that the value of $\Delta G^{\ddagger}$ is quite different from the experimental activation energy.

ACKNOWLEDGEMENTS

Grateful acknowledgement is made to Professor J. N. Murrell of Sussex University for providing details for the calculation of the molecular potential energy curves and surfaces of Figures 29 and 30; Figure 13 was supplied by the Royal Society of Chemistry; Figure 21 is based on a diagram in *Physical Chemistry* (1978) by P. W. Atkins, Oxford University Press; Figure 24 is based on a diagram from *Reaction Kinetics* (1975), Oxford Chemistry Series, no. 22, Oxford University Press; the data in Table 13 come from *Chemical Kinetics: A Modern Survey of Gas Reactions* (1976) by J. E. Nicholas, Harper & Row. The cover photograph was supplied by Lawrence Berkeley Laboratory/Science Photo Library.